Lecture Notes in Mathematics 1916

Fraydoun Rezakhanlou · Cédric Villani

Entropy Methods
for the Boltzmann Equation

Lectures from a Special Semester
at the Centre Émile Borel,
Institut H. Poincaré, Paris, 2001

Editors:

François Golse
Stefano Olla

 Springer

Authors and Editors

Fraydoun Rezakhanlou
Department of Mathematics
Evans Hall
University of California
Berkeley, CA 94720-3840, USA
e-mail: rezakhan@Math.Berkeley.edu

Cédric Villani
Unité de mathématiques
  pures et appliquées
(UMR CNRS 5669)
Ecole Normale Supérieure de Lyon
46, allée d'Italie
69364 Lyon Cedex 07, France
e-mail: Cedric.Villani@umpa.ens-lyon.fr

François Golse
Laboratoire Jacques-Louis Lions
(UMR CNRS 7598)
Université Pierre et Marie Curie
175 rue du Chevaleret
75013 Paris, France
e-mail: golse@math.jussieu.fr

Stefano Olla
Centre de recherche en mathématiques
  de la décision
(UMR CNRS 7534)
Université Paris - Dauphine
Place du Maréchal de Lattre de Tassigny
75775 Paris Cedex 16, France
e-mail: olla@ceremade.dauphine.fr

Library of Congress Control Number: 2007932803

Mathematics Subject Classification (2000): 76P05, 82B40, 94A15, 60K35, 82C22

ISSN print edition: 0075-8434
ISSN electronic edition: 1617-9692
ISBN 978-3-540-73704-9 Springer Berlin Heidelberg New York
DOI 10.1007/978-3-540-73705-6

Springer is a part of Springer Science+Business Media
springer.com
© Springer-Verlag Berlin Heidelberg 2008

Typesetting by the authors and SPi using a Springer LaTeX macro package

Cover design: *design & production* GmbH, Heidelberg

Printed on acid-free paper   SPIN: 12092147   41/SPi   5 4 3 2 1 0

Foreword

One of the major contributions of Ludwig Boltzmann to science has been the connection between time irreversibility and the increase of entropy as a well-defined quantity associated to the macroscopic state of a system. His ideas are at the basis of most studies in non-equilibrium statistical mechanics, and many non-equilibrium processes are still now physically understood in terms of their *entropy production.*

More recently entropy and entropy production have become mathematical tools used in the context of kinetic and hydrodynamic limits, when deriving the macroscopic behavior of systems from the interaction dynamics of their (many) microscopic elementary constituents at the atomic or molecular level.

In this volume, we have put together two surveys on some recent results in this direction. The first text, by Cedric Villani, illustrates the use of entropy in the analysis of convergence to equilibrium for solutions of the Boltzmann equation. The second text, by Fraydoun Rezakhanlou, discusses the Boltzmann–Grad limit, in which the Boltzmann equation is derived from the dynamics of a large number of hard spheres. Both entropy and entropy production play a major role in these problems.

To illustrate the relevance of entropy in both the kinetic theory of gases and the dynamics of a large number N of hard spheres, we shall recall below two fairly classical, and yet fundamental properties of Boltzmann's entropy.

The first property, which is a particular case of the Gibbs principle, is a variational characterization of Maxwellian equilibrium distributions in the kinetic theory of gases. Call $f \equiv f(x, v)$ the single-particle phase space density of molecules that are located at the position x with velocity v. (Here, both x and v run through the Euclidian space $\mathbf{R}^3$ for simplicity.) The following mechanical observables are easily defined in terms of the density f (and the mass m of each molecule):

$$\iint_{\mathbf{R}^3 \times \mathbf{R}^3} f(x, v)\mathrm{d}x\,\mathrm{d}v = \text{number of molecules},$$

$$\iint_{\mathbf{R}^3 \times \mathbf{R}^3} mvf(x, v)\mathrm{d}x\,\mathrm{d}v = \text{total momentum},$$

$$\iint_{\mathbf{R}^3 \times \mathbf{R}^3} \tfrac{1}{2}m|v|^2 f(x, v)\mathrm{d}x\,\mathrm{d}v = \text{total energy}.$$

Boltzmann's notion of entropy defined in terms of the density f is $-H(f)$, where the functional $H(f)$ is defined as

$$H(f) = \iint_{\mathbf{R}^3 \times \mathbf{R}^3} f(x, v)\ln f(x, v)\mathrm{d}x\,\mathrm{d}v\,.$$

Consider the following minimization problem:

$$\inf H(f) \text{ with constraints } \iint_{\mathbf{R}^3 \times \mathbf{R}^3} \begin{pmatrix} 1 \\ mv \\ \tfrac{1}{2}m|v|^2 \end{pmatrix} f(x, v)\mathrm{d}x\,\mathrm{d}v = \begin{pmatrix} N \\ P \\ E \end{pmatrix},$$

where $N, E \geq 0$ and $P \in \mathbf{R}^3$ are given.

There are obvious compatibility conditions to be verified by N, P, E for the set of functions f satisfying the constraints to be non-empty: for instance, by the Cauchy–Schwartz inequality, one should have

$$|P|^2 \leq 2mNE\,.$$

Forgetting momentarily the obvious constraint $f \geq 0$, we write the Euler equation for this minimization problem as

$$\begin{aligned}
\mathrm{D}H(f) \cdot \delta f &= \iint_{\mathbf{R}^3 \times \mathbf{R}^3} (\ln f(x, v) + 1)\delta f(x, v)\mathrm{d}x\,\mathrm{d}v \\
&= a \iint_{\mathbf{R}^3 \times \mathbf{R}^3} \delta f(x, v)\mathrm{d}x\,\mathrm{d}v \\
&\quad + b \iint_{\mathbf{R}^3 \times \mathbf{R}^3} mv\delta f(x, v)\mathrm{d}x\,\mathrm{d}v \\
&\quad + c \iint_{\mathbf{R}^3 \times \mathbf{R}^3} \tfrac{1}{2}m|v|^2 \delta f(x, v)\mathrm{d}x\,\mathrm{d}v\,,
\end{aligned}$$

where $a, c \in \mathbf{R}$ and $b \in \mathbf{R}^3$ are the Lagrange multipliers associated to the constraints of total number of molecules, total energy and total momentum. Since this equality must hold for each smooth, compactly supported δf, it follows that:

$$\ln f(x, v) + 1 = a + b \cdot (mv) + c\tfrac{1}{2}m|v|^2\,,$$

or, in other words,

$$f(x, v) = \mathrm{e}^{(a-1)+b\cdot(mv)+c\frac{1}{2}m|v|^2}\,.$$

Notice that the minimizing function f so defined is positive: we therefore verify a posteriori that there was no need for any Lagrange multiplier associated with the constraint $f \geq 0$ a.e..

This expression can be put in the more familiar form of a Maxwellian density

$$f(x, v) = \frac{N}{(2\pi k\theta)^{3/2}} e^{-m|v-u|^2/2k\theta}$$

by putting

$$a = 1 + \ln N - \tfrac{3}{2}\ln(2k\theta) - \frac{m|u|^2}{2k\theta}, \quad b = \tfrac{1}{k\theta}u, \quad c = -\tfrac{1}{k\theta},$$

where k is the Boltzmann constant. The bulk velocity u and temperature θ are related to the total momentum P and the total energy E by the formulas

$$P = Nmu \quad \text{and} \quad E = N(\tfrac{1}{2}m|u|^2 + \tfrac{3}{2}k\theta).$$

This computation shows that Maxwellian distributions are the critical points of the Boltzmann entropy on the affine manifold of densities f corresponding to a prescribed total number of molecules, total momentum and total energy.

By the strict convexity of the map $f \mapsto f \ln f$, one easily concludes that this critical point is in fact a global minimum of H.

To summarize: Maxwellian distributions maximize the Boltzmann entropy under the constraints of a fixed total number of molecules, total momentum and total energy.

Besides, the strict convexity of H implies that the relative entropy

$$H(f) - \inf H$$

defines some kind of distance from f to the set of Maxwellian distributions.

The second property of the entropy which we want to discuss is a variational characterization of chaotic densities. Let $F \equiv F(z_1, \ldots, z_N)$ be the N-body phase space probability density of a system of particles. Here, $z_i = (x_i, v_i)$ consists of the position x_i and velocity v_i of the ith particle; obviously z_i runs through $\mathbf{R}^3 \times \mathbf{R}^3$. We denote $Z_N = (z_1, \ldots, z_N)$ and $\hat{Z}_N^i = (z_1, \ldots, z_{i-1}, z_{i+1}, \ldots, z_N)$. To the density F is associated the family of its marginals

$$F_i(z_i) = \int_{(\mathbf{R}^3 \times \mathbf{R}^3)^{N-1}} F(Z_N) \mathrm{d}\hat{Z}_N^i, \quad i = 1, \ldots, N.$$

Consider the following minimization problem: to find

$$\inf H(F) \quad \text{under the constraints } F_i = f$$

where f is a given a.e. non-negative function in $L^1(\mathbf{R}^3 \times \mathbf{R}^3)$. Since

$$F \geq 0 \text{ a.e. and } \int_{(\mathbf{R}^3 \times \mathbf{R}^3)^N} F(Z_N) \mathrm{d}Z_N = 1,$$

the function f should satisfy

$$f \geq 0 \text{ a.e. and } \int_{\mathbf{R}^3 \times \mathbf{R}^3} f(z)\mathrm{d}z = 1$$

in order for the set of constraints to define a non-empty set of probability densities F.

Neglecting again the obvious constraint $F \geq 0$ a.e., we write the Euler equation for the minimization problem above as

$$\mathrm{DH}(F) \cdot \delta F = \int_{(\mathbf{R}^3 \times \mathbf{R}^3)^N} (\ln F(Z_N) + 1)\delta F(Z_N)\mathrm{d}Z_N$$
$$= \sum_{1 \leq i \leq N} \int_{(\mathbf{R}^3 \times \mathbf{R}^3)^N} a_i(z_i)\delta F(Z_N)\mathrm{d}Z_N \,.$$

Since this equality must be satisfied by each smooth, compactly supported δF, one must have

$$\ln F(Z_N) + 1 = \sum_{1 \leq i \leq N} a_i(z_i)$$

i.e.

$$F(z_1, \ldots, z_N) = \exp\left(\sum_{i=1}^{N} a(z_i) - 1\right) \,.$$

In other words, F is of the form

$$F(z_1, \ldots, z_N) = \prod_{i=1}^{N} \phi_i(z_i) \,, \quad \text{with } \phi_i = \exp(a_i - \tfrac{1}{N}) \,.$$

Writing

$$F(z_1, \ldots, z_N) = \prod_{i=1}^{N} \int_{\mathbf{R}^3 \times \mathbf{R}^3} \phi_i(z)\mathrm{d}z \prod_{i=1}^{N} \psi_i(z_i)$$

with

$$\psi_i = \frac{\phi_i}{\displaystyle\int_{\mathbf{R}^3 \times \mathbf{R}^3} \phi_i(z)\mathrm{d}z} \,,$$

we see that, on account of the normalization condition on F, one has

$$\prod_{i=1}^{N} \int_{\mathbf{R}^3 \times \mathbf{R}^3} \phi_i(z)\mathrm{d}z = 1$$

and

$$\psi_i = f \text{ for each } i = 1, \ldots, N \,.$$

In other words, the only critical point F for this minimization problem is the chaotic density

$$F(z_1, \ldots, z_N) = \prod_{i=1}^{N} f(z_i).$$

Hence, chaotic densities are the only critical points of the Boltzmann entropy on the affine manifold of probability densities with all their marginals equal to a given probability density f.

By using again the strict convexity of the functional H, we see that this critical point is in fact the minimum point.

Therefore, chaotic densities maximize the Boltzmann entropy among all probability densities with all their marginals equal to a given probability density f.

Again, the strict convexity of H implies that the relative entropy

$$H(F) - \inf H$$

measures the distance from F to the set of chaotic distributions.

The two properties of the entropy described above pertain to the two topics addressed in this volume.

Indeed, the first text, by Cédric Villani, concerns the use of the entropy and entropy production as a tool in order to estimate the speed of convergence to a (uniform) Maxwellian equilibrium density. As was explained above, the relative entropy measures the *distance* of a non-equilibrium state to equilibrium; entropy production is another way to measure that distance. Finding how these two measures of the distance to equilibrium are related is one of the major arguments in estimating the speed of approach to equilibrium in the kinetic theory of gases.

In spatially inhomogeneous non-equilibrium states it is then useful to work with *local entropy* and *local entropy production*. Cédric Villani carefully explains the mathematical difficulties arising in this problem: the system can be locally close to equilibrium, and have small total entropy production, while still being far from the set of *global* equilibria.

In Villani's own words, "local equilibrium states are your worst enemies" if you want to prove (and estimate) convergence to global equilibrium. In order to obtain this global convergence, the system should locally move out of the "local equilibrium" and Cédric Villani discusses various tools and conjectures on this mostly open problem.

This convergence problem is another aspect that shows the inadequacy of the notion of local equilibrium in order to understand non-equilibrium phenomena. Also transport in stationary non-equilibrium states (like the heat conductivity when the system is under a gradient of temperature imposed by external thermostats) cannot be explained in terms of local equilibrium states. Such local equilibrium states can only be a zeroth-order approximation of the real non-equilibrium state, and only further order approximations can explain transport and convergence to equilibrium.

The Boltzmann–Grad limit is the process by which the Boltzmann equation, which governs the evolution of the single-particle phase space density of the molecules of a monatomic gas, is derived from the N-body molecular dynamics. Hence, this limit necessarily involves the approximation of the N-body phase space density by chaotic densities whose single-body marginal is a solution of the Boltzmann equation. Therefore, the second property of entropy recalled above obviously plays a role in this limit.

The text by Fraydoun Rezakhanlou discusses various aspects of the Boltzmann–Grad limit. This is a classical open problem in mathematical physics, where little progress has been made since the seminal work of Lanford in 1975 (extended by Illner and Pulvirenti in 1986).

Rezakhanlou recalls the main conjecture, that can be formulated as a law of large numbers in a non-equilibrium situation, and also formulates the corresponding conjectures about small and large fluctuations about this limit. Then he propose a stochastic version of the hard sphere dynamics. Stochasticity helps in proving molecular chaos (the *Stosszahlansatz*) which is the key argument in all derivations of this type of limits.

We hope that these surveys, addressing two very different issues in the statistical mechanics of non-equilibrium processes with similar methods based on the concept of entropy as defined by Boltzmann, will convince the reader of the versatility of that notion.

We conclude this brief overview with a few words about the origin of these texts. In the fall term of 2001, we organized a four-month session supported by the Centre Émile Borel on "Hydrodynamic Limits" at the Institut Henri Poincaré in Paris. Various events were proposed in this period, including an international congress focussed on the state of the art as well as open problems and perspectives in the subject of hydrodynamic limits. This congress was dedicated to Claude Bardos in recognition of his fundamental contributions to this subject. In addition, several research courses were given during that period, among these the courses by Cedric Villani and Fraydoun Rezakhanlou whose notes are gathered together in this volume.

We express our deepest gratitude to both directors of Institut Henri Poincaré, Profs. Michel Broué and Alain Comtet, for the warm hospitality so generously offered to all participants in this session.

Our heartfelt thanks also go to all members of the staff at the Institute Henri Poincaré for their most competent help throughout the organization of this session.

Finally, the tragic events of September 11 2001 regretably struck the family of one of our guests; we are especially grateful to Mrs Annie Touchant and Mrs. Sylvie Lhermitte of the Centre Emile Borel for their kind assistance and support in these sad circumstances.

Paris, December 2006   François Golse
                                                                   Stefano Olla

Contents

Chapter 1
Entropy Production and Convergence to Equilibrium

C. Villani

Abstract This set of notes was used to complement my short course on the convergence to equilibrium for the Boltzmann equation, given at Institut Henri Poincaré in November–December 2001, as part of the *Hydrodynamic limits* program organized by Stefano Olla and François Golse. The informal style is in accordance with the fact that this is neither a reference book nor a research paper. The reader can use my review paper, *A review of mathematical topics in collisional kinetic theory*, as a reference source to dissipate any ambiguity with respect to notation for instance. Apart from minor corrections here and there, the main changes with respect to the original version of the notes were the addition of a final section to present some more recent developments and open directions, and the change of the sign convention for the entropy, to agree with physical tradition. Irene Mazzella is warmly thanked for kindly typesetting a preliminary version of this manuscript.

1.1 The Entropy Production Problem for the Boltzmann Equation

I shall start with Boltzmann's brilliant discovery that the H functional (or negative of the entropy) associated with a dilute gas is nonincreasing with time. To explain the meaning of this statement, let me first recall the model used by Boltzmann.

1.1.1 The Boltzmann Equation: Notation and Preliminaries

Unknown. $f(t, x, v) = f_t(x, v) \geq 0$ is a time-dependent probability distribution on the phase space $\Omega_x \times \mathbb{R}_v^N$, where $\Omega_x \subset \mathbb{R}^N$ ($N = 2$ or 3) is the

C. Villani
ENS Lyon, France e-mail: cvillani@umpa.ens-lyon.fr

spatial domain where particles evolve and $\mathbb{R}_v^N$ is the space of velocities (to be thought of as a tangent space).

Evolution equation.

$$
\begin{cases}
\dfrac{\partial f}{\partial t} + v.\nabla_x f = Q(f,f) \\
\qquad\qquad := \displaystyle\int_{S^{N-1}} \int_{\mathbb{R}^N} (f'f'_* - ff_*)B(v-v_*,\sigma)\,dv_*\,d\sigma
\end{cases}
\tag{BE}
$$

+ boundary conditions.

Notation.

- $f = f(t,x,v),\ f' = f(t,x,v'),\ f_* = f(t,x,v_*),\ f'_* = f(t,x,v'_*),$
- $v' = \dfrac{v+v_*}{2} + \dfrac{|v-v_*|}{2}\sigma,\quad v'_* = \dfrac{v+v_*}{2} - \dfrac{|v-v_*|}{2}\sigma \qquad (\sigma \in S^{N-1})$

Think of (v', v'_*) as possible pre-collisional velocities in a process of elastic collision between two particles, leading to post-collisional velocities $(v, v_*) \in \mathbb{R}^N \times \mathbb{R}^N$.

Physical quantity. $B = B(v - v_*, \sigma) \geq 0$, the collision kernel ($=$ cross-section times relative velocity) keeps track of the microscopic interaction. It is assumed to depend only on $|v - v_*|$ and $\cos\theta$, where

$$
\cos\theta = \left\langle \frac{v - v_*}{|v - v_*|}, \sigma \right\rangle.
$$

(Brackets stand for scalar product.) By abuse of notation I may sometimes write $B(v - v_*, \sigma) = B(|v - v_*|, \cos\theta)$.

The picture of collisions is as follows (in $\mathbb{R}_v^N$):

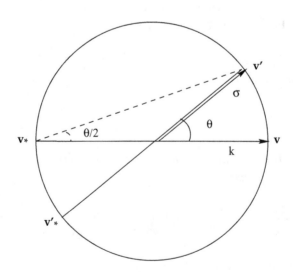

Boundary conditions. I shall consider three simple cases:

(1) Periodic condition: $\Omega = \mathbb{T}^N$ (not really a subset of $\mathbb{R}^N$!), no boundaries.
(2) Bounce-back condition: Ω smooth bounded,

$$f_t(x, v) = f_t(x, -v) \quad \text{for } x \in \partial\Omega$$

(3) Specular reflection: Ω smooth bounded,

$$f_t(x, R_x v) = f_t(x, v) \quad \text{for } x \in \partial\Omega,$$

$$\text{where } \begin{cases} R_x v = v - 2\langle n(x), v\rangle n(x) \\ n(x) = \text{normal to } \partial\Omega \quad \text{at } x \end{cases}$$

Local hydrodynamic fields. The following definitions constitute the bridge between the kinetic theory of Maxwell and Boltzmann on one hand, and the classical hydrodynamics on the other hand. Whenever $f(x, v)$ is a kinetic distribution, define the

- Local density $\rho(x) = \displaystyle\int_{\mathbb{R}_v^N} f(x, v)\, dv$
- Local velocity (mean) $u(x) = \dfrac{\int f(x, v)\, v\, dv}{\rho(x)}$
- Local temperature $T(x) = \dfrac{\int f(x, v)|v - u(x)|^2 dv}{N\rho(x)}$

A simple symmetry argument shows that $\displaystyle\int_{\mathbb{R}^N} Q(f, f)\varphi\, dv = 0$ for $\varphi = \varphi(v)$ in $\mathrm{Vect}\,(1, v_i, |v|^2)_{1 \le i \le N}$, as soon as $f = f(v)$ is integrable enough at large velocities. Those φ's are called **collision invariants**.

Global conservation laws. Let $(f_t)_{t \ge 0}$ be a well-behaved solution of the BE. Then

$$\begin{cases} \dfrac{d}{dt} \displaystyle\int f_t(x, v)\, dv\, dx = 0 \quad \text{(conservation of mass)} \\[4mm] \dfrac{d}{dt} \displaystyle\int f_t(x, v)\dfrac{|v|^2}{2}\, dv\, dx = 0 \quad \text{(conservation of kinetic energy)} \end{cases}$$

- Also $\dfrac{d}{dt} \displaystyle\int f_t(x, v)v\, dv\, dx = 0$ in the case of periodic boundary conditions (conservation of momentum).
- When Ω has an axis of symmetry $\mathbf{k}$ and specular reflection is enforced, then there is an additional conservation law:

$$\frac{d}{dt} \int f_t(x, v)\, v_0\, (k{\wedge}n)\, dv\, dx = 0 \text{ (conservation of angular momentum)}$$

($|k| = 1$, and $n = n(x)$ is still the normal).

Normalizations. Without loss of generality I shall assume

- $\displaystyle\int f_t(x,v)\,\mathrm{d}v\mathrm{d}x = 1 \qquad \int f_t(x,v)\frac{|v|^2}{2}\,\mathrm{d}v\,\mathrm{d}x = \frac{N}{2}$
- $\displaystyle\int f_t(x,v)\,v\,\mathrm{d}v\,\mathrm{d}x = 0$ in the periodic case
- $|\Omega| = 1$ ($|\Omega| = N$-dimensional Lebesgue measure of Ω)

Moreover, in this course I will *not* consider the case when Ω has an axis of symmetry and specular boundary condition is imposed. A discussion would have to take into account angular momentum, and consider separately the particular case when Ω is a ball.

1.1.2 H Functional and H Theorem

Let us now introduce *Boltzmann's H functional*: when f is a probability distribution on $\Omega \times \mathbb{R}^N$, define

$$H(f) = \int f \log f \, \mathrm{d}v\,\mathrm{d}x.$$

This quantity is well-defined in $\mathbb{R} \cup \{+\infty\}$ provided that $\displaystyle\int f(x,v)|v|^2\,\mathrm{d}v\,\mathrm{d}x$ is finite, and will be identified with the negative of the entropy associated with f.

The following theorem, essentially due to Boltzmann, will be our starting point.

Theorem 1. *Let $(f_t)_{t\geq 0}$ be a well-behaved (smooth) solution of the BE (in particular with finite entropy), with one of the boundary conditions discussed above. Then*

(i) $\displaystyle\frac{\mathrm{d}}{\mathrm{d}t}H(f_t) \leq 0$. *Moreover, one can define a functional D on $L^1(\mathbb{R}_v^N)$, called "entropy production functional", or "dissipation of H functional", such that*

$$\frac{\mathrm{d}}{\mathrm{d}t}H(f_t) = -\int_{\Omega_x} D(f_t(x,.))\,\mathrm{d}x.$$

(ii) Assume that the collision kernel $B(v - v_*, \sigma)$ is > 0 for almost all $(v, v_*, \sigma) \in \mathbb{R}^{2N} \times S^{N-1}$. Let $f(x,v)$ be a probability distribution distribution on $\Omega \times \mathbb{R}^N$, with $\displaystyle\int f(x,v)|v|^2\,\mathrm{d}v\,\mathrm{d}x < +\infty$. Then

$$\int_\Omega D(f(x,.))\,\mathrm{d}x = 0 \Longleftrightarrow f \text{ is in local equilibrium, i.e. there exist}$$

$$\text{functions } \rho(x) \geq 0, \ u(x) \in \mathbb{R}^N, \ T(x) > 0,$$

$$\text{such that } f(x,v) = \rho(x)\frac{e^{-|v-u(x)|^2/2T(x)}}{[2\pi T(x)]^{N/2}}.$$

(iii) Assume that the boundary condition is either periodic, or bounce-back, or specular, and in the latter case assume that the dimension is either 2 or 3 and that Ω has no axis of symmetry (is not a disk or a cylinder or an annulus or a ball or a shell). Without loss of generality, assume that f satisfies the normalizations discussed above. Then

$$(f_t)_{t\geq 0} \text{ is stationary} \iff \forall t \geq 0 \quad \int_\Omega D(f_t(x,.))dx = 0$$

$$\iff f_t(x,v) = \frac{e^{-|v|^2/2}}{(2\pi)^{N/2}} \equiv M(v) \quad \forall t \geq 0$$

The proof of this theorem is well-known (actually there are several proofs for point (ii), even though not so many), but is is useful to sketch it in order to help understanding refinements to come.

Proof of Theorem 1 (sketch).

(i)

$$\frac{d}{dt} \int f_t \log f_t = \int Q(f_t, f_t)(\log f_t + 1) - \int (v.\nabla_x f_t)(\log f_t + 1)$$

$$= \int Q(f_t, f_t) \log f_t - \int \nabla_x \cdot (v f_t \log f_x)$$

$$= \int_{\Omega \times \mathbb{R}^N} Q(f_t, f_t) \log f_t - \int_{\partial\Omega \times \mathbb{R}^N} [v \cdot n(x)] f_t \log f_t$$

Under any one of the boundary conditions that we use, the second integral is 0. As for the first one, it can be rewritten as

$$\int_\Omega \int_{\mathbb{R}^{2N}} \int_{S^{N-1}} (f' f'_* - f f_*) \log f \, B(v - v_*, \sigma) \, dv \, dv_* \, d\sigma \, dx$$

By a simple symmetry trick, this is also

$$-\frac{1}{4} \int_\Omega \int_{\mathbb{R}^{2N}} \int_{S^{N-1}} (f' f'_* - f f_*) \log \frac{f' f'_*}{f f_*} B \, d\sigma \, dv_* \, dv \, dx$$

which takes the form $- \int_\Omega D(f) \, dx$ if one defines the *entropy production functional*:

$$D(f) = \frac{1}{4} \int_{\mathbb{R}^{2N} \times S^{N-1}} (f' f'_* - f f_*) \log \frac{f' f'_*}{f f_*} B(v - v_*, \sigma) \, d\sigma \, dv_* \, dv.$$

Clearly $D(f) \geq 0$ because $B \geq 0$ and $(X-Y) \log \frac{X}{Y} \geq 0$ as a consequence of log being increasing.

(ii) Since $B > 0$ almost everywhere, the equality means that for (almost) all $x \in \Omega$ the L^1 function $f = f(x, \cdot)$ satisfies the functional equation of Maxwell–Boltzmann:

$$f(v')f(v'_*) = f(v)f(v_*) \quad \text{for almost all } v, v_*, \sigma \tag{MB}$$

(and also $\int f(v)(1 + |v|^2) \, dv < +\infty$, up to deletion of a negligible set of x's).

Integrate equation (MB) with respect to $\sigma \in S^{N-1}$, to find that

$$f(v)f(v_*) = \frac{1}{|S^{N-1}|} \int_{S^{N-1}} f(v')f(v'_*) \, d\sigma$$

$$= \frac{1}{|S(v, v_*)|} \int_{S(v, v_*)} f(\alpha)f(\tilde{\alpha}) \, d\alpha$$

where

- $S(v, v_*)$ is the collision sphere, centered at $\dfrac{v + v_*}{2}$, with radius $\dfrac{|v - v_*|}{2}$
- $\tilde{\alpha}$ is the symmetric of α with respect to $\dfrac{v + v_*}{2}$.

The important point about this average over $S(v, v_*)$ is that it only depends upon $S(v, v_*)$, whence only upon $\dfrac{v + v_*}{2}$ and $\dfrac{|v - v_*|}{2}$, or (which is equivalent) upon the physically meaningful variables

$$\begin{cases} m = v + v_* \quad \text{(total momentum)} \\[2mm] e = \dfrac{|v|^2 + |v_*|^2}{2} \quad \text{(total kinetic energy)} \end{cases}$$

Thus $f(v)f(v_*) = G(m, e)$. *Note.* In this argument, due to Boltzmann, the Maxwell distribution arises from this *conflict of symmetries* between the tensor product structure of ff_* and the dependence of g upon a reduced set of variables: m and e.

Let us continue with the proof of (ii). We first assume f to be smooth (C^1, positive). Taking logarithms, we find

$$\log f(v) + \log f(v_*) = \log G(m, e)$$

$$\frac{\partial}{\partial v} \implies \nabla \log f(v) + 0 = \nabla_v \Big[\log G(m, e) \Big]$$

$$= \nabla_m \Big[\log G(m, e) \Big] + \frac{\partial}{\partial e} \Big[\log G(m, e) \Big] v$$

Similarly, $\nabla \log f(v_*) = \nabla_m \Big[\log G(m, e) \Big] + \dfrac{\partial}{\partial e} \Big[\log G(m, e) \Big] v_*$. So $(\nabla \log f)(v) - (\nabla \log f)(v_*) \mathbin{/\!/} v - v_* \quad \forall v, v_* \in \mathbb{R}^N \times \mathbb{R}^N$.

As a purely algebraic consequence of this (here $N \geq 2$) is crucial), there exists $\lambda \in \mathbb{R}$ and $\mu \in \mathbb{R}^N$ such that

$$\forall v \in \mathbb{R}^N \quad \nabla \log f(v) = \lambda v + \mu.$$

This in turn implies that f is a Maxwellian distribution.

What happens if f is not smooth? As remarked by Desvillettes, the Maxwell–Boltzmann equation (MB), written with a suitable parametrization of the collisions, is invariant under convolution with a Maxwellian. In fact the following property will be enough for us: if

$$M_\delta(v, v_*) = M_\delta(v) M_\delta(v_*) = \frac{e^{-|v|^2/2\delta}}{(2\pi\delta)^{N/2}} \frac{e^{-|v_*|^2/2\delta}}{(2\pi\delta)^{N/2}}$$

then from $f(v)f(v_*) = G(m, e)$ we may deduce

$$M_\delta \underset{v, v_*}{*} \quad ff_* \equiv M_\delta \underset{v, v_*}{*} \quad G(m, e)$$

which is $(f * M_\delta)(f * M_\delta)_* = M_\delta * G(m, e)$. But the convolution by M_δ preserves the class of functions of the form $G(v + v_*, |v - v_*|)$.

Hence, for all $\delta > 0$, $(f * M_\delta)(f * M_\delta)_* = G^\delta(m, e)$, so $f * M_\delta$ is a Maxwellian distribution, and we conclude by letting $\delta \to 0$ that f also is.

Remark. Lions [56] has a beautiful direct proof that (MB) $\Rightarrow$ [f is C^∞].

Now let us go on with the proof of Theorem 1 (iii). It is clear from (ii) that $Q(M, M) = 0$. Also $v \cdot \nabla_x M = 0$, so M is a stationary solution. Conversely, let $(f_t)_{t \geq 0}$ be a solution which does not produce entropy. From (ii) we know that

$$f_t(x, v) = \rho(t, x) \frac{e^{-(|v - u(t,x)|^2)/2T(t,x)}}{[2\pi T(t, x)]^{N/2}}.$$

So we can plug this into the BE, which reduces to $(\partial f / \partial t) + v \cdot \nabla_x f = 0$ since $Q(f_t, f_t) = 0$. Assuming f smooth > 0, we write

$$\frac{1}{f} \left(\frac{\partial f}{\partial t} + v \cdot \nabla_x f \right) = 0$$

and see that it is in fact a system of polynomial equations in v:

$$\frac{1}{\rho} \frac{\partial \rho}{\partial t} + \frac{|v - u|^2}{2T^2} \frac{\partial T}{\partial t} + \left\langle \frac{v - u}{T}, \frac{\partial u}{\partial t} \right\rangle - \frac{N}{2T} \frac{\partial T}{\partial t}$$

$$+ \frac{v \cdot \nabla_x \rho}{\rho} + \frac{|v - u|^2}{2T^2} v \cdot \nabla_x T - \frac{N}{2T} v \cdot \nabla_x \rho + \frac{\nabla_x u}{T} : [v \otimes (v - u)] = 0.$$

Let us use the shorthand $X' = \partial X / \partial t$, and identify powers of v:

$$(0) \quad \frac{\rho'}{\rho} + \frac{|u|^2}{2T^2} T' - \frac{N}{2T} T' - \frac{\langle u, u' \rangle}{T} = 0$$

(1) $\quad -\dfrac{T'}{T^2}u + \dfrac{u'}{T} + \dfrac{\nabla\rho}{\rho} + \dfrac{|u|^2}{2T^2}\nabla T - \dfrac{N}{2T}\nabla T - \dfrac{1}{T}\nabla\left(\dfrac{|u|^2}{2}\right) = 0$

(2) $\quad \dfrac{T'}{2T^2}\delta_{ij} - \dfrac{u_j\partial_i T + u_i\partial_j T}{2T^2} + \dfrac{\partial_i u_j + \partial_j u_i}{2T} = 0$

(3) $\quad \dfrac{\nabla T}{2T^2} = 0$

For a detailed version of the reasoning which follows, see [36].

- From (3) we have $\nabla_x T = 0$, so $T = T(t)$.
- Plugging this into (2) we find

$$\frac{\partial_i u_j + \partial_j u_i}{2} = -\frac{T'}{2T}\delta_{ij} \quad \text{(in matrix notation, } \nabla_{\text{sym}} u = -\frac{T'}{2T}\,\text{Id})$$

and this expression is independent of x. From this one easily sees that $\partial_{ij} u_i = 0$, then $\partial_{ii} u_j = 0$, then u is an affine map which can be written as $u(x) = \lambda x + \Lambda \wedge x + u_0$. ($\lambda \in \mathbb{R}$, $\Lambda \in \mathbb{R}^3$)

From the boundary conditions, $\displaystyle\int_\Omega (\nabla \cdot u)\,\mathrm{d}x = \int_{\partial\Omega} u \cdot n = 0$ (periodic $\Rightarrow \partial\Omega = \emptyset$, bounce-back $\Rightarrow u = 0$ on $\partial\Omega$, specular $\Rightarrow u \cdot n = 0$). So $\lambda = 0$ and $u(x) = \Lambda \wedge x + u_0$. This is possible for $\Lambda \neq 0$ only if Ω has an axis of symmetry $k \parallel \lambda$ and we are considering specular reflection, a case which we have excluded. So $u(x) = u_0(t)$, but u_0 can be $\neq 0$ only in the periodic case. To sum up, at this point we know that $u \equiv 0$; or $u \equiv u(t)$ in the periodic case.

- Going back to (1): if we know $u \equiv 0$, then

$$\frac{\nabla\rho}{\rho} = 0, \quad \text{so } \rho = \rho(t);$$

and if we know $u = u(t)$, then $\nabla\rho/\rho$ is independent of x, so $\rho = \exp(A(t) \cdot x + B(t)) = \rho_0(t)e^{A_0(t)\cdot x}$.

- Then, from (0) ρ' only depends on t, and so does ρ.

Conclusion. In all the cases, we know that:

- ρ, u, T do not depend on x (case of the torus)
- or ρ, T do not depend on x and $u \equiv 0$ (other cases).

But the global conservation laws imply $\displaystyle\int \rho = 1$, $\displaystyle\int \rho\frac{|u|^2}{2} + \frac{N}{2}\int \rho T = \frac{N}{2}$ (conservation of mass and energy). Also in the periodic case $\displaystyle\int \rho u = 0$ (conservation of momentum). This implies $\rho \equiv 1$, $u \equiv 0$, $T \equiv 1$. The proof is complete. $\square$

Theorem 1 is very important! Point (ii) is at the basis of the problem of hydrodynamic limit, while point (iii) is crucial in the problem of trend to equilibrium. We should note that in the particular situation when $f_t(x, v) = f_t(v)$ [spatial homogeneity, $x \in \mathbb{T}^N$] then (iii) is superfluous.

1.1.3 What this Course is About: Convergence to Equilibrium

Problem of trend to equilibrium. Let be given a solution $(f_t)_{t \geq 0}$ of the BE, starting from some initial datum f_0 which is out of equilibrium (i.e., with our normalizations, $f_0 \neq M$). Is it true that

$$f_t \underset{t \to \infty}{\longrightarrow} M?$$

Remark 1. A major difference between the problem of hydrodynamic limit and the problem of trend to equilibrium is that in the latter case, one is interested in proving that the solution approaches a *global* equilibrium, while in the former one only expects to come close to a *local* equilibrium. Accordingly, as a general rule the problem of trend to equilibrium is much more sensitive to boundary conditions, than the hydrodynamic limit problem.

Remark 2. Another difference is that one expects shocks occurring asymptotically as the Knudsen number goes to O, at least in the compressible hydrodynamic limit, while for fixed Knudsen number there is no known reason for appearance of shocks – at least in a strictly convex domain.

Remark 3. The H Theorem is *not* the only way to attack the problem of trend to equilibrium. But it seems to be by far the most robust.

Remark 4. A lot of other kinetic (or not) models have similar features. The choice of the BE is interesting because of historical reasons, because it is one of the very, very few models which have been derived from Newton's laws of motion "alone" (in some particular regimes), and also because it concentrates a lot of serious difficulties.

As soon as one is able to get a little bit of information about solution of the BE (which is very difficult!), one can turn Theorem 1 into some statement of trend to equilibrium. For instance we have the following result by some easy compactness argument:

Soft Theorem: Let $(f_t)_{t \geq 0}$ be a solution (meaning not made precise here) of the BE, with $H(f_0) < +\infty$ and

$$\sup_{t \geq 0} \int f_t(x, v) |v|^{2+\delta} \, dv \, dx < +\infty \quad \text{for some} \ \delta > 0$$

then

$$f_t \underset{t \to \infty}{\longrightarrow} M \ \text{in weak} - L^1 \ \text{sense}$$

Let me insist that nobody knows how to construct such solutions under natural assumptions on f_0 (say, f_0 belonging to Schwartz class of C^∞ rapidly decaying functions)!! Only under some restrictions on f_0 are such a priori estimates available. But this is a different issue, it concerns the study of the Cauchy problem, not so much the mechanism of trend to equilibrium.

Goal of the course: Present a program on entropy production, indirectly inspired by the pioneering works of Kac (in the 1950s) and McKean (in the 1960s), really started by Carlen and Carvalho at the beginning of the 1990s, currently very active. Its goal is to *provide explicit bounds on the rate of convergence by a detailed analysis of the entropy production in situations where the Cauchy problem is well understood.*

When Kac and McKean began to think about convergence to equilibrium, the models they were able to "attack" were extremely simplified compared to those that experts in the Cauchy problem were able to treat. This remained so for a long time, but now it is the reverse situation: the study of the Cauchy problem, while very active, is still very far from completion, while the entropy production program, likely, will soon cover all interesting situations under assumptions of suitable a priori estimates (exactly which estimates is an interesting point because this may be taken as a guide for the Cauchy problem).

The recent story of entropy production has been very much influenced by the field of *logarithmic Sobolev inequalities* (but as we shall see, there are a lot of distinctive, very interesting features) and by some theoretical portions of *information theory*.

The basic idea behind the entropy production program is quite natural and simple: given that (f_t) is far from equilibrium at some time t, try to estimate from below the amount of entropy which has to be produced at later times.

Here is an example of a recent result (based on work by Mouhot and myself) obtained by this program:

Let $f_0(v)$ satisfy $\int f_0(v)\,dv = 1$, $\int f_0(v)\,v\,dv = 0$, $\int f_0(v)\frac{|v|^2}{2}\,dv = \frac{N}{2}$,

$\int f_0(v)\,|v|^3\,dv < +\infty$, $f_0 \in L^p$ for some $p > 1$. Let $(f_t)_{t\geq 0}$ be the unique solution of the spatially homogeneous Boltzmann equation (SHBE) starting from f_0 with the collision kernel $B(v - v_*, \sigma) = |v - v_*|$ (in dimension $N = 3$ this is the usual hard spheres collision kernel). Then

$$\left\| f_t - M \right\|_{L^1} = O(t^{-\infty}).$$

More precisely, for any $\epsilon > 0$ there exists an explicit constant $C_\epsilon(f_0)$ such that $\forall t > 0$ $\|f_t - M\|_{L^1} \leq C_\epsilon t^{-1/\epsilon}$.

Nota. Since the first construction of solutions of the SHBE with hard sphere cross-section by Carleman [16], this is the very first result asserting better decay than just $O(t^{-K})$ for some $K > 0$ [79] with explicit constants.

I shall conclude this introduction with two remarks.

Remark 5. As I said above, the program was intended to cover situations where the Cauchy problem is well understood. Does it provide information in more general situations? I don't know! The reader can try to extract from the arguments some principles which would help in such settings.

Remark 6. In principle the entropy production program can also be applied to hydrodynamic limits. This was in fact the main motivation of Carlen and Carvalho [19, 20]. Carlen et al. [25] have demonstrated on an interesting baby model (for which the hydrodynamic model is a set of two nonlinear ODE's) that this goal is sometimes feasible. But there is in fact not much in common between these two problems, besides the fact that tools which are used to estimate entropy production in a certain context, are likely to be useful to estimate entropy production in another context.

1.2 Tentative Panomara

The entropy production program is closely linked to several other topics which possess their own interest:

- The central limit theorem for Maxwellian molecules: this is an alternative approach of trend to equilibrium, which has been first suggested by McKean [60], developing on an idea by Kac. Just as the central limit theorem of i.i.d. random variables, the central limit theorem for Maxwellian molecules was implemented quantitatively by several methods, in particular via:
  - Fourier transform (Toscani and co-authors)
  - Monge–Kantorovich distances (Tanaka)
  - Information-theoretical techniques (Carlen and Carvalho)

  (McKean's method was a clever variant of Trotter's elementary proof of the central limit theorem). This program has recently led to very sharp results for the study of trend to equilibrium, but it seems to concern *only* the case where there is no x (position) variable and in addition $B(|v - v_*|, \cos \theta) = b(\cos \theta)$ [spatially homogeneous Maxwellian case].
- Kac's problem of trend to equilibrium for an N-particle system: this is an attempt of understanding trend to equilibrium on the basis of the asymptotic (as $N \to \infty$ and $t \to \infty$ together) properties of the master equation describing the evolution of a system of N colliding particles. After decades of sleep, this program has been revived by recent works (Janvresse; Carlen, Carvalho and Loss; Maslen), but it is still very far from satisfactory understanding.
- The behavior of Fisher's information along solutions of the Boltzmann equation: McKean [60] made the striking discovery that the Fisher information $I(f) = \int \dfrac{|\nabla f|^2}{f}$ is a Lyapunov functional for a simple one-dimensional caricature of the Boltzmann equation. He believed this was a particular feature of dimension 1, but it is not, and this result has been extended to multi-dimensional Boltzmann equations (Toscani and subsequent work), however, only in the case of "Maxwellian molecules". There is

a tight connection between this problem and the entropy production issue, as we shall see.

I shall briefly review these three topics in Sect. 1.6. Some information can also be found in my 2003 paper about "Cercignani's conjecture". Also, there are alternative programs based on entropy production:

- Compactness arguments (possibly for weak solutions), non-quantitative.
  $\longrightarrow$ In the context of hydrodynamic limits, compactness is crucial for the Bardos–Golse–Levermore program. The text of my 2001 Bourbaki seminar (published in 2002) tries to describe the essentials of that program, which was recently led to completion by the joint efforts of Bardos, Golse, Levermore, Lions, Masmoudi, Saint-Raymond.
  $\longrightarrow$ In the problem of trend to equilibrium, compactness arguments are rather easy and general but quite disappointing: at present, they can essentially be applied only in situations where quantitative arguments lead to infinitely stronger results (and yet, not all of them!). This is because the bound on $\int f_t(x,v)|v|^{2+\delta}\,\mathrm{d}v\,\mathrm{d}x$ is very difficult to establish.
- Linearization in a close-to-equilibrium setting. In this process $H(f)$ turns into the $(L^2\text{ norm})^2$ of the perturbation $f - M$ in some weighted L^2 space: $\int \dfrac{(f-M)^2}{M}\,\mathrm{d}v\,\mathrm{d}x$ (extremely strong norm!). After a change of unknown, trend to equilibrium can hopefully be studied by spectral gap methods. This approach is unbeatable in a (very-)close-to-equilibrium setting, but *cannot* predict anything quantitative when f_0 is not very close to M.

On the next page is an attempt to summarize the whole picture.

1.3 Reminders from Information Theory

In this section I shall review some necessary preliminaries before turning to the study of the Boltzmann equation.

1.3.1 Background and Definitions

Information theory was founded by Shannon [69] for applications to signal transmission, error-correcting, etc. An excellent reference about modern information theory is Cover and Thomas [32], *Elements of information theory*. For the particular topics which will be discussed here, another user-friendly reference is Chap. 10 of the recent book *Sur les inégalités de Sobolev logarithmiques* [2] (in French). Part of information theory deals with the study of functionals which aim at quantifying the "information" contained in

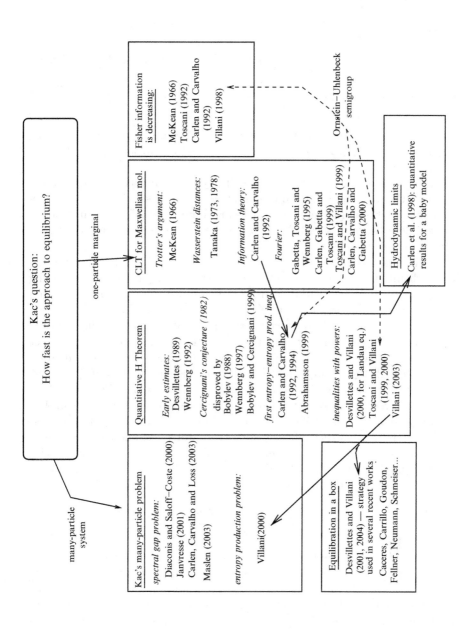

a random variable; the two most popular of these functionals were introduced respectively by Fisher [43] and Shannon [69]:

$$\text{Shannon's entropy } S(f) = -\int_{\mathbb{R}^N} f \, \log \, f$$
$$= \text{Boltzmann's } H \text{ function, up to change of sign}$$

$$\text{Fisher's information } I(f) = \int_{\mathbb{R}^N} \frac{|\nabla f|^2}{f} = 4 \int_{\mathbb{R}^N} |\nabla \sqrt{f}|^2 = -\int_{\mathbb{R}^N} f \Delta(\log f).$$

Whenever X is a random variable whose law has Lebesgue density f, one defines $S(X) = S(f)$, $I(X) = I(f)$. In the sequel I shall work with $H = -S$ instead of S.

Historical motivations. In Shannon's theory of communication, the entropy measures the (asymptotically) optimal rate of compression that one can apply to a signal without (asymptotically) losing any of the information which it carries. Here "asymptotically" means: in the limit when one cuts the signal into *very long* samples.

In Fisher's theory of "efficient statistics", Fisher's information measures how difficult it is to infer the value of an unknown parameter (appearing in the law of a random variable) from the empirical observation. Typically: assume that you observe a signal $X_\theta = X + \theta$, where the law of X is known and equal to f, and $\theta \in \mathbb{R}$ (say) is deterministic but unknown to the observer. Given a large sample of independent observations of X_θ, one can construct a statistic $\hat{\theta}$ (observed values) which is hoped to give a good approximation of θ, with high probability. But the Cramér–Rao inequality tells you that, no matter how cleverly you design $\hat{\theta}$, if it is unbiased (i.e. $E\hat{\theta} = \theta$, where E stands for expectation), then its variance cannot be smaller than $1 \, / \, [nI(f)]$ (here $n =$ number of observations). Hence, the lower $I(f)$ is, the most difficult it is to construct efficient statistics. Moreover, one can show that, asymptotically as $n \to \infty$, the maximum likelihood estimate achieves this lower bound $[nI(f)]^{-1}$. In higher dimensions ($N \geq 1$), the Fisher information *matrix* is a basic object in statistics:

$$I_{ij}(f) = \int_{\mathbb{R}^N} \frac{\partial_i f \partial_j f}{f},$$

of which $I(f)$ is just the trace.

Remark. Fisher's information is sometimes encountered in kinetic theory under the name "Linnik functional", for the following reason: McKean [60] borrowed from Linnik [55] the idea to use Fisher's information in the context of the central limit theorem, and in his paper omitted to quote Fisher. This terminology should be banned to avoid further confusion.

A first hint that H and I can be used to represent some kind of "information" is their super-additive properties: an information about the joint law of (X,Y) is usually much more valuable than an information about the marginal laws of X and Y. And, as a matter of fact, it is the case that

$$H((X,Y)) \geq H(X) + H(Y)$$
$$I((X,Y)) \geq I(X) + I(Y),$$

and in both cases, equality occurs if and only if X, Y are independent.

The first inequality (superadditivity of H) is well-known (see any textbook on information theory). The second one can be proven by a simple and short computation, which the reader can try to find out as an exercise; it is also a particular case of more general and more delicate information-theoretical inequalities proven by Carlen [17, 18].

Conventions. In a more intrinsic definition, H and I should be defined relatively to some *reference measure* ν (not necessarily a probability):

$$H_\nu(f) = \int f \log f \, d\nu, \quad I_\nu(f) = \int \frac{|\nabla f|^2}{f} \, d\nu.$$

I shall use the notation

$$H(\mu|\nu) = H_\nu\left(\frac{d\mu}{d\nu}\right), \qquad I(\mu|\nu) = I_\nu\left(\frac{d\mu}{d\nu}\right)$$

and call these functionals the relative entropy of μ with respect to ν, relative Fisher information of μ with respect to ν. In information theory $H(\mu|\nu)$ is called the Kullback entropy of μ with respect to ν, or informational divergence, or Kullback–Leibler distance, and probably has many other names.

Abuse of notation. When $d\mu = f(x)\,dx$, $d\nu = g(x)\,dx$, I shall write $H(f|g)$, $I(f|g)$ for $H(\mu|\nu)$, $I(\mu|\nu)$.

Finally, an object which is very popular in information theory, ever since the time of Shannon, is the *power entropy*:

$$\mathcal{N}(f) = \frac{e^{-\frac{2}{N}H(f)}}{2\pi e} = \mathcal{N}(X) \quad \text{if law } (X) = f(x)\,dx$$

1.3.2 Inequalities

We can now state three famous inequalities about H and I.

1. *The Blachman–Stam inequality* (three *equivalent* formulations)

$$I(\sqrt{\lambda}X + \sqrt{1-\lambda}Y) \leq \lambda I(X) + (1-\lambda)I(Y)$$
$$I(X+Y) \leq \lambda^2 I(X) + (1-\lambda)^2 I(Y)$$
$$\frac{1}{I(X+Y)} \geq \frac{1}{I(X)} + \frac{1}{I(Y)},$$

whenever $\lambda \in [0,1]$ and X, Y are independent random variables.

2. *The Shannon–Stam inequality* (three equivalent formulations)

$$H(\sqrt{\lambda}X + \sqrt{1-\lambda}Y) \leq \lambda H(X) + (1-\lambda)H(Y)$$

$$H(X+Y) \leq \lambda H(X) + (1-\lambda)H(Y)$$

$$+\frac{N}{2}[\lambda \log \lambda + (1-\lambda)\log(1-\lambda)]$$

$$\mathcal{N}(X+Y) \geq \mathcal{N}(X) + \mathcal{N}(Y)$$

3. *The Stam–Gross logarithmic Sobolev inequality* (three equivalent formulations)

$$\begin{cases} H(f \mid M) \leq \dfrac{1}{2}I(f \mid M) \\[2ex] \mathcal{N}(f)I(f) \geq N \end{cases}$$

whenever f is a probability density on $\mathbb{R}^N$

$$\int h^2 \log h^2 M \leq 2 \int |\nabla h|^2 M + \left(\int h^2 M \right) \log \left(\int h^2 M \right)$$

Paternity issues. Shannon [69] conjectured inequality 2 (in the $\mathcal{N}$ formulation) and found it useful for certain issues in signal transmission. Then Stam [71] proved 1 and 2, using a suggestion by de Bruijn. In the same work he noticed the validity of 3 (in the second formulation) for $N = 1$ (actually, the general case can be reduced to this particular case). Later Blachman [7] gave an implied proof for 1. In 1975 Gross, unaware of Stam's work, proved 3 (in the third formulation) and, most importantly, showed that this inequality is equivalent to the hypercontractivity of the Ornstein–Uhlenbeck semigroup. (Although this is not widely acknowledged, the concept of hypercontractivity goes back at last to Bonami [13].) A few years later, Lieb [54], unaware of Shannon and Stam, re-proved 2 as a consequence of optimal Young inequalities. It is only at the beginning of the nineties that the whole picture was clarified, with the papers of Carlen [17], providing synthetical and simple proofs of all this, together with multiple links.

I myself wrote a very short paper about the entropy power functional [84].

De Bruijn's identity. Let $(P_t)_{t\geq 0}$ be the heat semigroup (generated by the PDE $\partial_t f = \Delta f$), then, regardless of regularity issues (which do not really bother here),

$$\frac{\mathrm{d}}{\mathrm{d}t}H(P_t f) = -I(P_t f).$$

It is easy to check this fact, which as we shall see is very convenient.

Geometrical interpretation. There is an appealing interpretation of all this in geometrical language. From Boltzmann's original definition of entropy

("$S = k \ln W$") we know that $S = -H$ can be thought of as an infinite-dimensional analogue of $A \mapsto \dfrac{1}{n} \log |A|_n$ ($A \subset \mathbb{R}^n$, $|A|_n$ = Lebesgue measure of A), or equivalently that $\mathcal{N}$ plays the role of $A \mapsto |A|_n^{2/n}$. Under this analogy, one can develop the following correspondence:

$$\text{probability measures } f \rightarrow \text{convex sets } A \text{ in } \mathbb{R}^n \ (n \to \infty)$$

$$\text{entropy power } \mathcal{N}(f) \rightarrow |A|_n^{1/N}$$

$$\text{Maxwellian distribution } M \rightarrow \text{ball of radius } \sqrt{n} \text{ in } \mathbb{R}^n, B_n(\sqrt{n})$$

$$\text{Shannon–Stam ineq.} \rightarrow \text{Brunn} - \text{Minkowski inequality}$$

$$|A + B|_n^{1/n} \geq |A|_n^{1/n} + |B|_n^{1/n}$$

$$(\text{or rather the restricted version}:$$

$$|A + B|_n^{2/n} \geq |A|_n^{2/n} + |B|_n^{2/n})$$

$$\text{Logarithmic Sobolev ineq.} \rightarrow \text{isoperimetric inequality}$$

$$\frac{|\partial A|_{n-1}}{n|A|_n}|A|_n^{1/n} \geq |B_n(1)|_n^{1/n}$$

$$\text{Fisher information } I(f) \rightarrow \frac{\text{surface } (A)}{n|A|_n} = \frac{|\partial A|_{n-1}}{n|A|_n}.$$

In this array there is just one missing inequality for the analogy to be complete: this is the Blachman–Stam inequality. The following finite-dimensional analogue of it was suggested by Dembo et al. [34]:

$$\frac{|(A + B)|_n}{|\partial(A + B)|_{n-1}} \geq \frac{|A|_n}{|\partial A|_{n-1}}$$

Here $|\partial A|_{n-1}$ is the surface of A, defined as $\lim\limits_{\epsilon \to 0} \dfrac{|A + \epsilon B_n(1)|_n - (A)}{\epsilon}$.

This inequality is apparently not known. It cannot hold time for general compact sets A, B, but it has a chance to be true for convex bodies, and by means of some Alexandrov–Fenchel inequalities for mixed volumes, one can check that it *is* true if A is convex and B is a ball.

Before going back to Boltzmann's H theorem, let us record three particular formulations of the above-mentioned inequalities; these are the ones which one should mostly keep in mind for applications in kinetic theory. Let

$$f_\lambda = \text{law of } \sqrt{\lambda}X \quad \text{if } X \text{ is random with law } f(x)\,dx.$$

Then

$$
\begin{cases}
I(f_{\frac{1}{2}} * g_{\frac{1}{2}}) \leq \dfrac{1}{2}\Big[I(f) + I(g)\Big] \\[3mm]
H(f_{\frac{1}{2}} * g_{\frac{1}{2}}) \leq \dfrac{1}{2}\Big[H(f) + H(g)\Big] \\[3mm]
H(f \mid M) \leq \dfrac{1}{2}I(f \mid M).
\end{cases}
$$

1.4 Quantitative H Theorem

1.4.1 Entropy–Entropy Production Inequalities

Recall part (ii) of Theorem 1: whenever $B > 0$ a.e. and $f \in L_2^1(\mathbb{R}_v^N)$ $(f \geq 0)$ satisfies $D(f) = 0$ [B enters in the expression of D, so an assumption on B is mandatory], then f is a Maxwellian distribution:

$$
f(v) = \rho \, \frac{e^{-|v-u|^2/2T}}{(2\pi T)^{N/2}}. \tag{1}
$$

We now wish to get a quantitative replacement for that, stating that if B is positive enough and $D(f)$ is small, then f is close to a Maxwellian. The first such results were established by Desvillettes [35] in the form (essentially)

$$
D(f) \geq K_{f,R} \inf_{m \in \mathcal{M}} \int_{|v|<R} |\log f - \log m| \, dv \tag{2}
$$

where $\mathcal{M}$ is the space of all Maxwellian distributions, and $K_{f,R}$ depends on R and f via certain bounds, in particular on the strict positivity of f. Desvillettes' theorem was established under the assumption that B is bounded from below, but later Wennberg [89] relaxed this restriction.

Essentially the best that one can hope for is an *entropy-entropy production inequality*, which would relate $D(f)$ on one hand, $H(f \mid M^f)$ on the other hand, where M^f is the Maxwellian distribution having the same hydrodynamic parameters (mass, momentum, energy) as f. In the present context, an *EEP* inequality is a functional inequality of the form

$$
D(f) \geq \Theta\Big(H(f \mid M^f)\Big). \tag{3}
$$

[Note that $H(f \mid M^f) = H(f) - H(M^f)$], where $\Theta : \mathbb{R}_+ \to \mathbb{R}_+$ is some increasing function, with $\Theta(\alpha) > 0$ if $\alpha > 0$. Such an inequality implies at once convergence to equilibrium *for the spatially homogeneous BE* (only collisions are taken into account, there is no dependence on the position).

Moreover, one can then estimate the speed of convergence to equilibrium in term of the behavior of $\Theta(\alpha)$ close to $\alpha = 0$; the most favorable situation being that when $\Theta(\alpha) = K \cdot \alpha$ for α small enough (exponential decay).

We already encountered an EEP inequality for another model: consider the (linear) *Fokker–Planck equation*

$$\frac{\partial f}{\partial t} = \Delta f + \nabla \cdot (fv) \quad t \geq 0, \ v \in \mathbb{R}^N \tag{4}$$

For this equation the role of the (negative of the) entropy is played by the Kullback information $H(f \mid M)$, and its dissipation along the Fokker–Planck semigroup is

$$D_{\mathrm{FP}}(f) = I(f \mid M)$$

Thus the *logarithmic Sobolev inequality* of Sect. 3 states that, for any probability distribution f,

$$D_{\mathrm{FP}}(f) \geq 2H(f \mid M) \tag{5}$$

which incidentally implies by Gronwall's lemma $H(S_t f \mid M) \leq \mathrm{e}^{-2t} H(f \mid M)$ $S_t = $ semigroup associated to the FP equation (4). The assumption that f is a probability distribution is of course not the point: an adequate change of parameters allows to treat any nonnegative distribution.

Inequality (5) is a *very strong* EEP inequality: first because the function Θ is linear ($\Theta(\alpha) = 2\alpha$), second because it holds true without restriction upon f. For more complicated models, certainly we will not be so lucky and Θ may be vanishing to higher order, or the inequality may hold only under some restrictions (smoothness, decay, positivity, etc.).

1.4.2 EEP Inequalities for the Boltzmann Equation

The goal of this chapter is to discuss EEP inequalities for the Boltzmann collision operator, under various assumptions on the kernel B and on f.

Normalization. Assume

$$\int_{\mathbb{R}^N} f \, dv = 1, \quad \int_{\mathbb{R}^N} fv \, dv = 0, \quad \int_{\mathbb{R}^N} f|v|^2 \, dv = N. \tag{6}$$

Then

$$M^f = M = \frac{\mathrm{e}^{-|v|^2/2}}{(2\pi)^{N/2}}.$$

Notation. This class of f's will be denoted by $\mathcal{C}(1,0,1)$.

It is easy to extend the results which will be established under this normalization, to the general case. Indeed, if f has mass ρ, momentum u, temperature T, then

$$f(v) = \frac{\rho}{T^{N/2}}\left(\frac{v-u}{\sqrt{T}}\right), \quad \text{where} \quad \int \tilde{f}(v) \begin{pmatrix} 1 \\ v \\ |v|^2 \end{pmatrix} dv = \begin{pmatrix} 1 \\ 0 \\ N \end{pmatrix}$$

and one has

$$\begin{cases} H(f \mid M^f) = \rho\, H(\tilde{f} \mid M) \\ D(f) \qquad = \rho^2\, \tilde{D}(\tilde{f}) \end{cases}$$

where $\tilde{D}$ is the entropy production functional associated with the modified kernel $\tilde{B}(v - v_*, \sigma) = B(\sqrt{T}(v - v_*), \sigma)$.

It was once conjectured [31] that under suitable assumptions on B, there would exist a constant $K > 0$ such that for all f satisfying (6), $D(f) \geq KH(f \mid M)$. This conjecture was then disproved under more and more general assumptions [8, 9, 96]. In particular, the following important "counter-theorem" by Bobylev and Cercignani [9] shows that there is no hope of such an inequality holding true, even if we restrict to probability densities f which are very smooth (in the sense of Sobolev norms), decay quite fast at infinity (in the sense of having a lot of finite moments), and are strictly positive (in the strong sense of being bounded below by a Maxwellian distribution). Before I state the theorem, some further remarks and notation:

Remark 1. I consider these estimates (Sobolev norms, moments, lower bound like $Ke^{-c|v|^q}$) as the natural estimates associated with the nonlinear Boltzmann equation. This will be discussed in more detail in the next chapter. I stress that, in a *linearized* context, it would be natural to look at stronger estimates like $L^p(M^{-1}dv)$ norms.

Functional spaces.

$$\|f\|_{L^p_s} = \left[\int_{\mathbb{R}^N} f(v)^p \left(1 + |v|^p\right)^s dv\right]^{1/p};$$

in particular $\|f\|_{L^1_s} = \int f(v)(1 + |v|)^s dv;$

$$\|f\|_{H^k} = \left[\sum_{j \leq k} \int_{\mathbb{R}^N} |D^j f(v)|^2 dv\right]^{1/2}.$$

In the sequel I shall use L^1_s and H^k norms as a natural way to quantify the decay at infinity and the smoothness, respectively. It would also be possible to combine both informations in a single two-parameter family of norms:

$$\|f\|_{H^k_s} = \left[\int_{\mathbb{R}^N} |D^k f(v)|^2 (1 + |v|^2)^s dv\right]^{1/2},$$

but let's rather consider separately decay and smoothness. At last, here comes the above-mentioned theorem (this is no exactly the formulation given in the reference quoted, but it is straightforward to make the transformation):

Theorem 2 (Bobylev & Cercignani, 1999). *Let B satisfy*

$$\int_{S^{N-1}} B(v - v_*, \sigma)\mathrm{d}\sigma \leq C_B(1 + |v - v_*|^\gamma)$$

for some $\gamma \in [0, 2]$. Then there exist sequences $(M_s)_{s \geq 0}$, $(S_k)_{k \geq 0}$ [to be understood as moment and smoothness bounds] and ρ_0, $A_0 > 0$, such that

$$\inf \left\{ \begin{array}{c} \dfrac{D(f)}{H(f \mid M)}; \ f \in \mathcal{C}(1, 0, 1); \quad \forall k \in I\!N \ \|f\|_{H^k} \leq S_k; \\[2mm] \forall s \in I\!N \ \|f\|_{L_s^1} \leq M_s; \quad \forall v \in I\!R^N \ f(v) \geq \rho_0 e^{-A_0|v|^2} \end{array} \right\} = 0.$$

In particular, it is impossible to have an inequality like $D(f) \geq KH(f \mid M)$ if we only assume bounds on $\|f\|_{L_s^1}$, $\|f\|_{H^k}$, and a lower bound.

Remark 2. One can moreover impose that any finite number of the bounds M_s, S_k, ρ_0, A_0, be arbitrarily close to their equilibrium values. More precisely, for any s_*, k_*, and any $\epsilon > 0$, one can impose

$$\left\{ \begin{array}{l} M_s \leq \|M\|_{L_s^1} + \epsilon \quad \forall s \leq s_* \\[2mm] S_k \leq \|M\|_{H^k} + \epsilon \quad \forall k \leq k_* \\[2mm] \rho_0 \geq 1 - \epsilon \\[2mm] A_0 \leq \dfrac{1}{2} + \epsilon. \end{array} \right.$$

Principle of the proof. It consists in constructing an explicit sequence of functions f_n) with $\dfrac{D(f_n)}{H(f_n \mid M)} \xrightarrow[n \to \infty]{} 0$. The sequence is simply defined as a perturbation of M, with "fat tails" (in a very weak sense):

$$f_n = M + \epsilon_n \, \varphi\left(\frac{\cdot}{\theta_n}\right),$$

where θ_n is large enough, ϵ_n is small enough, and φ is an *arbitrary* smooth, positive function bounded below by a Maxwellian.

Then f_n does not exactly belong to $\mathcal{C}_{1,0,1}$, but a slight perturbation of it (multiply by a constant and rescale slightly) does. The important point is that θ_n is very large, so that $\varphi(v/\theta_n)$ decays quite "slowly" at infinity, but ϵ_n is very, very small, so that f_n really is a perturbation of the equilibrium.

Next come some positive results. The first EEP inequality for the Boltzmann operator was established by Carlen and Carvalho (1992), [19]. This a very clever, complicated construction, and the function Θ appearing in the EEP inequality was very, very flat close to 0. A huge improvement was then provided by Toscani and Villani [79], and recently I improved again on these results [87]. In the sequel I shall describe a subset of what is presently known.

Theorem 3 ([78]). *Let B satisfy*

$$B(v - v_*, \sigma) \geq K_B(1 + |v - v_*|)^{-\beta} \quad \text{for some} \quad \beta \geq 0.$$

Then, for all $\epsilon > 0$,

$$\forall p > 1, \quad \forall I, \quad \forall \rho_0, \forall A_0, \forall q_0, \ \exists s = s(\epsilon, p, q_0 B); \quad \forall M$$

$$\exists K_\epsilon > 0; \quad \left[\|f\|_{L^p} \geq I, \quad \|f\|_{L^1_s} \leq M, \quad f \geq \rho_0 e^{-A_0 |v|^{q_0}} \right]$$

$$\implies D(f) \geq K_\epsilon H(f \mid M)^{1+\epsilon}.$$

In a more intelligible statement: $D(f) \geq K_\epsilon(f) H(f \mid M)^{1+\epsilon}$, where $K_\epsilon(f)$ only depends on an upper bound on $\|f\|_{L^p}$ ($p > 1$ arbitrary), an upper bound on high enough order moments of f, and a lower bound on f. (In short $K_\epsilon(f)$ depends on the integrability - decay - strict positivity of f.)

Shorthand. Let's just write "$D(f) \geq K(f) H(f \mid M)^{1+0}$ if f has some integrability, decays very fast, is strictly positive".

Theorem 4 ([87]). *Let B satisfy*

$$B(v - v_*, \sigma) \geq K_B \inf \left(|v - v_*|^\gamma, |v - v_*|^{-\beta} \right) \quad \text{for some} \quad \beta, \gamma \geq 0.$$

Then, for all $\epsilon > 0$

$$\forall \rho_0 \quad \forall A_0 \quad \forall q_0 \quad \exists k = k(\epsilon, q_0, B) \quad \exists s = s(\epsilon, q_0, B) \; / \; \forall S \; \forall M$$

$$\exists K_\epsilon > 0; \quad \left[\|f\|_{H^k} \leq Z, \ \|f\|_{L^1_s} \leq M, \ f \geq \rho_0 e^{-A_0 |v|^{q_0}} \right]$$

$$\implies D(f) \geq K_\epsilon H(f \mid M)^{1+\epsilon}$$

In words: This theorem extends the previous result to the case when B is allowed to vanish (ex: hard spheres, $B = |v - v_*|$), at the price of requiring high-order regularity (Sobolev norms at any order). This result can be rewritten as

$$D(f) \geq K_\epsilon(f) H(f \mid M)^{1+\epsilon},$$

where $K_\epsilon(f)$ only depends on an "upper bound" on the Sobolev-regularity for some large enough exponent, an upper bound on moments of f of order high enough and a lower bound on f. (In short K_ϵ depends on the smoothness – decay – strict positivity of f.)

Comment 1. Putting together Theorems 2–4 we see that EEP inequalities for the Boltzmann equation are very tricky: then is no linear EEP inequality (with $\Theta(\alpha) = K\alpha$), but there are EEP inequalities as soon as one allows the power to be > 1 ($\Theta(\alpha) = K\alpha^{1+\epsilon}$). Probably there are some refinements like $K\alpha \log \dfrac{1}{\alpha}$, but this is not so immediate. Further see Sect. 1.4.3.

Comment 2. At the level of the spatially homogeneous Boltzmann equation,

$$
\left\{\begin{array}{l} \text{functional inequality} \\ D(f) \geq K H(f \,|\, M)^{1+\epsilon} \end{array}\right\} \underset{\substack{\text{a priori} \\ \text{estimates}}}{\Longrightarrow} \left\{\begin{array}{l} \text{estimate of rate of convergence} \\ H(f_t \,|\, M) = O(t^{-\infty}) \end{array}\right\}
$$

Here $O(t^{-\infty})$ means $O(t^{-K})$, for any $K > 0$. I shall come back to this in the next chapter.

Comment 3. As explained in the next chapter, too, the stronger assumptions in Theorem 4 (Sobolev regularity) will not necessarily result in stronger assumptions at the level of the trend of equilibrium problem.

Comment 4. All constants K_ϵ appearing in this theorem are *explicit* – but a bit complicated! See [87] for more details.

In all the sequel of this chapter, I will try to describe the proofs of Theorems 3 and 4 in an intelligible way.

Principle of the proof of Theorem 3. It is an adaptation to this Boltzmann context of the celebrated (among those who know it!) proof of the Shannon–Stam inequality by Stam himself, together with some of the ingredients used in Boltzmann's proof of Theorem 1 (ii).

In short, the main idea is to introduce an *auxiliary diffusion semigroup* which is compatible with the important symmetries of the problem, and try to show that the *dissipation* of our functional $D(f)$, along this semigroup, is bounded below in terms of how far f is from equilibrium. Once we have this information, we can integrate along the semigroup, to recover some information on the value of $D(f)$ itself.

To illustrate this strategy, I shall first recall how the argument goes for the Shannon–Stam inequality. Stam (building on an idea by de Bruijn) used the *heat semigroup* $(P_t)_{t \geq 0}$ to link the Shannon–Stam inequality with the Blachman–Stam inequality thanks to:

(1) The identity $-\dfrac{\mathrm{d}}{\mathrm{d}t} H(P_t f) = I(P_t f)$

(2) The nice properties of the heat semigroup with respect to the rescaled covolution, i.e. the operation which to f and g (laws of X and Y, respectively) associate $f_\lambda * g_{1-\lambda}$, the law of $\sqrt{\lambda}\, X + \sqrt{1 - \lambda}\, Y$:

$$
\left[f_\lambda(x) = \frac{1}{\lambda^{N/2}} \, f\!\left(\frac{x}{\sqrt{\lambda}}\right) \right].
$$

The advantage of this strategy is that I is much simpler than H in several respects, in particular because it involves quadratic quantities. This makes it rather simple to prove the Blachman–Stam inequality by some Cauchy–Schwarz inequalities, then use this semigroup argument to deduce from this the Shannon–Stam inequality.

Remark. This principle is also underlying the Γ_2 calculus of Bakry and Emery, or many semigroup arguments arising in the theory of logarithmic Sobolev inequalities. Semigroup arguments were once very popular in certain circles dealing with probability theory, geometry and convex bodies, then were superseded by other methods, and currently seem to regain vitality again.

Here is how the Shannon–Stam inequality can be proven elegantly, thanks to a variation of this principle [29]. First note that it is equivalent to

$$H(f_\lambda * g_{1-\lambda} \mid M) \le \lambda H(f \mid M) + (1-\lambda)H(g \mid M)$$

Next introduce the *Fokker–Planck* (or adjoint Ornstein–Uhlenbeck) semigroup $(S_t)_{t\ge 0}$, defined by the solution of $\partial_t f = \Delta f + \nabla \cdot (fv)$. (Here v is the variable in $\mathbb{R}^N$.) This semigroup has the advantage over (P_t), to possess a steady state. Then one shows that

$$-\frac{\mathrm{d}}{\mathrm{d}t} H(S_t f \mid M) = I(S_t f \mid M)$$

$$-\frac{\mathrm{d}}{\mathrm{d}t} H\Big((S_t f)_\lambda * (S_t g_{1-\lambda}) \mid M\Big) = I\Big((S_t f)_\lambda * (S_t g)_{1-\lambda} \mid M\Big),$$

and therefore one has the representation formula

$$H\Big((S_t f)_g * (S_t g)_{1-\lambda} \mid M\Big) = \int_0^{+\infty} I\Big((S_t f)_\lambda * (S_t g)_{1-\lambda} \mid M\Big)\mathrm{d}t$$

$$\uparrow$$

$$\text{because } H(S_t h \mid M) \xrightarrow[t \to 0]{} 0$$

By Blachman–Stam inequality, this is

$$\le \lambda \int_0^{+\infty} I(S_t f \mid M)\mathrm{d}t + (1-\lambda) \int_0^{+\infty} I(S_t g \mid M)\mathrm{d}t$$

$$= \lambda H(f \mid M) + (1-\lambda)H(g \mid M).$$

In the case of the Boltzmann entropy production problem, this strategy will turn out to work also (which is somewhat unexpected), but will lead to much more complications (as we can expect).

The auxiliary semigroup will still be the Fokker–Planck semigroup; in fact it is natural to use a Gaussian semigroup in view of the symmetries of the Boltzmann operator. So essentially we shall try to estimate *the dissipation, by the Fokker–Planck semigroup, of the entropy production* (which itself is computed along the Boltzmann equation). This is not very natural from the physical point of view – all the more that we shall not apply the Fokker–Planck semigroup directly to the entropy production functional, but rather to a suitable lower bound. And the incredible thing is that, essentially, the object which will appear rather naturally in this process is another interesting object in kinetic theory: the *Landau entropy production functional*,

$$D_{\mathrm{L}}(f) = \int_{\mathbb{R}^{2N}} \Psi(|v - v_*|)ff_* \Big|\Pi_{(v-v_*)^\perp}\Big[\nabla(\log f) - \nabla_*(\log f)_*\Big]\Big|^2 \, \mathrm{d}v \, \mathrm{d}v_*.$$

Thus, in some sense the Landau entropy production functional appears as a kind of infinitesimal variation of the Boltzmann entropy production functional, along the Fokker–Planck semigroup.

Remark. The Landau equation reads

$$\frac{\partial f}{\partial t} + v \cdot \nabla_x f = Q_L(f, f)$$

$$= \nabla_v \cdot \left(\int_{\mathbb{R}^N} a(v - v_*)[f_* \nabla f - f(\nabla f)_*] \, dv_* \right),$$

where $a(v - v_*) = \Psi(|v - v_*|) \, \Pi_{(v - v_*)^\perp}$, and $\Pi_{z^\perp}$ is the orthogonal projection (in $\mathbb{R}^N$) onto the orthogonal hyperplane to z). This equation appears in the physics of charged (plasma) particles, *only* in the particular case $\Psi(|z|) = K/|z|$ ($K > 0$). From the mathematical point of view, it can be studied for various choices of Ψ, and for our purpose it is $\Psi(z) = |z|^2$ which shows up. This equation is related to the Boltzmann equation by the asymptotics of grazing collisions: for references and rigorous results see, e.g., [86], [1].

Proof. The goal is to estimate from *below*

$$D(f) = \frac{1}{4} \int \left(f'f'_* - ff_* \right) \log \frac{f'f'_*}{ff_*} B(v - v_*, \sigma) \, d\sigma \, dv \, dv_*$$

$$\geq \frac{K_B}{4} \int \left(f'f'_* - ff_* \right) \log \frac{f'f'_*}{ff_*} (1 + |v - v_*|)^{-\beta} \, d\sigma \, dv \, dv_*$$

by assumption,

where $\beta \geq 0$ ($\beta = 0$ would be enough here, but would not simplify things very much). I shall divide the argument into eight lemmas, from which the whole proof can be reconstructed fairly easily. Some of them are very simple and some other are quite tricky. □

Lemma 1 (Boltzmann's angular integration trick).

$$D(f) \geq \frac{K_B |S^{N-1}|}{4} \overline{D}(f), \quad \text{where}$$

$$\overline{D}(f) \equiv \int (ff_* - \int f'f'_* d\sigma) \log \frac{ff_*}{\int f'f'_* d\sigma} (1 + |v - v_*|)^{-\beta} dv dv_*$$

and I uses the notation $\fint \cdot \, d\sigma = \frac{1}{|S|^{N-1}} \int_{S^{N-1}} \cdot \, d\sigma$ *(average).*

This lemma is a direct consequence of Jensen's inequality and the *joint convexity* of the function $(X - Y) \log X/Y$ on $\mathbb{R}_+ \times \mathbb{R}_+$.

Let me recall that $\fint f'f'_* \, d\sigma$ is a function of $\begin{cases} m = v + v_* \\ e = \dfrac{|v|^2 + |v_*|^2}{2} \end{cases}$ only.

Let us write $\begin{cases} F(v,v_*) = f(v)f(v_*) \\ G(v,v_*) = \int f'f'_* \mathrm{d}\sigma = G(m,e). \end{cases}$

Lemma 2 (The semigroup $(S_t)_{t\geq 0}$ is compatible with the basic symmetries).

$$\forall t \geq 0 \quad \begin{cases} S_t F(v,v_*) = S_t(ff_*) = (S_t f)(S_t f)_* \\ S_t G(v,v_*) \quad \text{is still a function of } m \text{ and } e \text{ only.} \end{cases}$$

Also $\displaystyle \int S_t f \begin{pmatrix} 1 \\ v \\ |v|^2 \end{pmatrix} \mathrm{d}v = \begin{pmatrix} 1 \\ 0 \\ N \end{pmatrix}$.

Notation. I have used the same notation (S_t) for the FP semigroup acting on $L^1(\mathbb{R}^N)$ and acting on $L^1(\mathbb{R}^{2N}) = L^1(\mathbb{R}^N_v \times \mathbb{R}^N_{v_*})$. This lemma is a consequence of the fact that the FP semigroup acts as the rescaled convolution with M.

Remark. In fact there is a stronger identity:

$$S_t \left(\int_{S^{N-1}} f'f'_* \, \mathrm{d}\sigma \right) = \int_{S^{N-1}} (S_t f)'(S_t f)'_* \, \mathrm{d}\sigma$$

Notation. In the sequel, $X = (v,v_*) \in \mathbb{R}^{2N}$.

Lemma 3 (Nonlinear commutator formula). *Let $(S_t)_{t\geq 0}$ be any semigroup with generator $L = \Delta + a(X) \cdot \nabla + b(X)$, acting on $L^1(\mathbb{R}^d_X)$. Then*

$$\frac{\mathrm{d}}{\mathrm{d}t}\Big|_{t=0} \left[S_t\left((F-G)\log\frac{F}{G} \right) - (S_t F - S_t G)\log\frac{S_t F}{S_t G} \right] = \left| \frac{\nabla F}{F} - \frac{\nabla G}{G} \right|^2 (F+G).$$

The proof is by direct computation: for instance the object to compute can be written as $L[(F-G)\log F/G] - (LF - LG)\log(F/G) - (F-G)((LF/F) - (LG/G))$. The terms in $a \cdot \nabla$ disappear because they are differentiation operators, the terms in b disappear because $(F-G)\log(F/G)$ is homogeneous of degree 1.

Remark. This is a kind of "tensorial analogue" of the usual formula

$$\Delta\Phi(f) - \Phi'(f)\Delta f = \Phi''(f)|\nabla f|^2,$$

a "Γ_1 formula" for the Laplace operator.

The following lemma is crucial. Note that t below has nothing to do with the time in the Boltzmann equation!

Lemma 4 (Representation formula for $\overline{D}(f)$). *Let ψ be any function of X, F and G be any probability distributions on $\mathbb{R}^d_X$, then*

$$\int \psi(X)(F-G)\log\frac{F}{G}\,dX = \int_0^{+\infty}\left[\int \psi(X)\left|\frac{\nabla S_t F}{S_t F} - \frac{\nabla S_t G}{S_t G}\right|^2 (S_t F + S_t G)\,dX\right]dt$$

"main" contribution

$$-\int_0^{+\infty}\left[\int L^*\psi(X)(S_t F - S_t G)\log\frac{S_t F}{S_t G}\,dX\right]dt$$

"error".

Here $(S_t)_{t\geq 0}$ is again the Fokker–Planck semigroup on $\mathbb{R}_X^d$. This formula holds true under the assumption that F, G are integrable enough (enough moments), and that ψ does not increase faster than polynomially in X. Formally, this is an immediate consequence of Lemma 3, integration against ψ, and the identity

$$\int \psi(X)(F-G)\log\frac{F}{G}\,dX = \int_0^{+\infty}\int\left[\frac{d}{dt}(S_t F - S_t G)\log\frac{S_t F}{S_t G}\right]\psi(X)\,dX\,dt,$$

which itself comes from $(S_t F - S_t G)\log\dfrac{S_t F}{S_t G} \xrightarrow[t\to\infty]{} 0$ since $S_t F \xrightarrow[t\to\infty]{} M$. The technical point in establishing the formula rigorously consist in controlling the convergence as $t \to \infty$, and the continuity as $t \to 0$, of the map $t \mapsto \int (S_t F - S_t G)\log\dfrac{S_t F}{S_t G}\psi\,dX$.

Warning. ∇ here is with respect to $X = (v, v_*)$ in our application.

Lemma 5 (Symmetries). *Let P be the linear operator on $\mathbb{R}^{2N}$, depending on X, defined by*

$$P : [A, B] \mapsto \Pi_{(v-v_*)^\perp}[A - B]$$

(in fact $P \in L^\infty(\mathbb{R}_v^N \times \mathbb{R}_{v_*}^N; \mathcal{L}(\mathbb{R}^{2N}, \mathbb{R}^N)))$ ($\mathcal{L}$ = the set of linear mappings).

Then, as soon as G only depends on $m = v + v_*$, $e = (|v|^2 + |v_*|^2)/2$, one has

$$\left|\frac{\nabla F}{F} - \frac{\nabla G}{G}\right|^2 \geq \frac{1}{\|P\|^2}\left|P\left(\frac{\nabla F}{F}\right)\right|^2$$

$$= \frac{1}{4}\left|\frac{P\nabla F}{F}\right|^2.$$

Corollary 1.

$$\left|\frac{\nabla S_t F}{S_t F} - \frac{\nabla S_t G}{S_t G}\right|^2 S_t F \geq \frac{1}{4}\left|\Pi_{(v-v_*)^\perp}\left[\frac{\nabla S_t f}{f} - \left(\frac{\nabla S_t f}{f}\right)_*\right]\right|$$

The proof is elementary linear algebra, plus the key property that

$$\frac{\nabla_X F}{F} = \left[\frac{\nabla f}{f}, \left(\frac{\nabla f}{f}\right)_*\right] \quad \text{if } F = f f_*$$

$$\frac{\nabla_X G}{G} = \left[\frac{\nabla_m G}{G} + v\frac{\partial_e G}{\partial e}, \frac{\nabla_m G}{G} + v_*\frac{\partial G}{\partial e}\right]$$

together with the definition of P.

Remark. It is worthwhile noting that, at this particular point, we are able to implement the symmetry part of the proof of Theorem 1(ii), without having had to differentiate f – because the semigroup has done it for us! So what we have introduced here is the bounded operator P, instead of an unbounded operator of differentiation.

Lemma 6 (A strong form of EEP inequality for the Landau equation).

Let $f(v) \geq 0$ satisfy $\displaystyle\int_{\mathbb{R}^N} f(v) \begin{pmatrix} 1 \\ v \\ |v|^2 \end{pmatrix} dv = \begin{pmatrix} 1 \\ 0 \\ N \end{pmatrix}$.

Let $\displaystyle T_f = \inf_{e \in S^{N-1}} \int f(v)(v \cdot e)^2 dv$ *("minimum directional temperature") and*

$$D_L(f) = \frac{1}{2} \int_{\mathbb{R}^{2N}} \Psi(|v - v_*|) f f_* \left| \Pi_{(v-v_*)^\perp} \left[\nabla(\log f) - \nabla_*(\log f)_* \right] \right|^2 dv\, dv_*.$$

Then $\Psi(|v - v_*|) \leq K_\Psi |v - v_*|^2 \implies D_L(f) \geq K_\Psi \left(\dfrac{N-1}{2}\right) T_f I(f \mid M).$

Remark 1. The assumption on Ψ is rather unrealistic in the case of the Landau equation (but used in numerical simulations and some theoretical studies). As we already mentioned, the physical Ψ is $1/|z|$.

Remark 2. If we apply the logarithmic Sobolev inequality we conclude that $D_L(f) \geq \left[K_\Psi(N-1)T_f\right] H(f \mid M)$ which is an EEP for the Landau equation in this situation – notice that this one is "linear"!

This lemma is one of the key estimates in Desvillettes and Villani [39]. To prove it, one first reduces to the case when $\displaystyle\int f(v)v_i v_j = T_i \delta_{ij}$ by a change of orthonormal basis, then expands everything, taking advantage of the quadratic nature of things. One then applies the two inequalities

$$\sum_{ij} \int \left(|v|^2 \delta_{ij} - v_i v_j\right) \frac{\partial_i f \partial_j f}{f} \, dv \geq 0,$$

$$\sum_i \alpha_i \int \frac{(\partial_i f)^2}{f} dv \geq \sum_i \frac{\alpha_i}{T_i} \quad \text{(a variant of Heisenberg's inequality)}.$$

Remark 3. The estimate $T_f > 0$ means that all directions are represented in the support of f, which is crucial in the proof of

$$\frac{\nabla f}{f} - \left(\frac{\nabla f}{f}\right)_* // v - v_* \Longrightarrow f \text{ is Gaussian.}$$

Lemma 7 (Going from I back to H by the FP semigroup).

$$\int_0^{+\infty} I(S_t f \mid M)\,dt = H(f \mid M).$$

This lemma is well known.

Lemma 8 $((S_t)_{t\geq 0}$ is well adapted to the above-mentioned inequalities). *If $f \in \mathcal{C}(\bar{1}, 0, 1)$ (normalization of mass, momentum, energy), and satisfies assumptions of boundedness of all moments, some L^p estimates and some lower bound estimates, then*

$$T_{S_t f} \geq T_f \tag{1}$$

$$\|S_t F\|_{L_s^1} \leq C_s(\|F\|_{L_s^1} + 1) \tag{2}$$

$$\|S_t F\|_{L_s^2 \log L} \leq C_s(\|F\|_{L_s^1 \log L} + \|F\|_{L_{s+1}^1} + 1)$$

$$\text{[obtained by ODE techniques and use of LSI]}$$

$$H(S_t F \mid M) \xrightarrow[t \to \infty]{} 0 \text{ exponentially fast (LSI)}$$

$$\forall R \geq 1, \; \forall s \tag{3}$$

$$\int_{|X| \geq R} |X|^2 (F - G) \log \frac{F}{G}\,dX \leq \frac{C_s(f)}{R^s}$$

$$\int_0^{+\infty} \left[\int_{|X| \geq R} |X|^2 (S_t F - S_t G) \log \frac{S_t F}{S_t G}\,dX\right] dt \leq \frac{C_s(f)}{R^s}$$

This step is just technical: a lot of bounds by use of Cauchy–Schwarz, Chebyshev, elementary inequalities, etc..

Finally, one ties up Lemmas 1–8 to get the desired estimate (Theorem 3). A conflict arises in the implementation, between what suggests Lemma 4 (choose $\psi(X)$ close to a constant, so that $L^*\psi = \Delta\psi - X.\nabla\psi$ is small, and the "error" is indeed small), and what suggests Lemma 6 (choose $\psi(X)$ close to $\geq |v - v_*|^2$ to apply the estimate). This conflict is resolved schematically as in the picture.

In the argument, this leads to an error term for velocities $\geq R$, where ψ is not dominated by $(1 + |v - v_*|)^{-\beta}$. This error term is however very

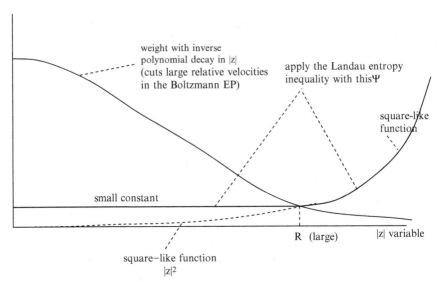

weight with inverse
polynomial decay in |z|
(cuts large relative velocities
in the Boltzmann EP)

apply the Landau entropy
inequality with this Ψ

square-like
function

small constant

R (large) |z| variable

square−like function
$|z|^2$

Fig. 1.1 Cooking up the truncation in the proof of Theorem 3

small, as a function of R, in view of Lemma 8. This small error term is the
reason why, in the end, we do not recover a bound below by $H(F \mid M)$, but
only by $[H(f \mid M)$-small error] (small term which can be chosen to be like
$H(f \mid M)^{1+\epsilon}$ for arbitrarily small ϵ), due to the fact that s in (3) of Lemma 8
is as large as one wishes – so error terms are $O(R^{-\infty})$.

All of this is implemented carefully in [79]. A technically simpler proof has
been later devised in [87]: in that variant, the error terms do not involve any
integration with respect to the semigroup time variable, which makes things
less intricate.

Once Theorem 3 is proven, Theorem 4 follows by a very tricky "interpolation" estimate: $\forall \epsilon > 0, \ \delta \leq 1$,

$$\int_{|v-v_*| \leq \delta} (f'f'_* - ff_*) \log \frac{f'f'_*}{ff_*} \, dv \, dv_* \, d\sigma \leq C_\epsilon(f) H(f \mid M)^{1-\epsilon} \delta^{1/4};$$

C_ϵ depends on moments, Sobolev norms and lower bound on f. The difficulty
here is to show that the left-hand side vanishes as $\delta \to 0$ (this is easy), but at
the same time as $f \to M$, and with an order very close to 2 (thing of $H(f \mid M)$
as a square norm). Il would not be sufficient to have just $O(H(f \mid M)^\alpha \delta^{1/4})$
for some $\alpha < 1$.

The bounds on the Sobolev norms of f enter the proof somewhere via the
interpolation inequality

$$\|u\|_{L^2} \leq C_N \|u\|_{L^1}^{1-\eta} \|u\|_{H^{\frac{N+1}{\eta}}}^{\eta}.$$

I won't say more about the argument because it is "purely technical": see [87].

1.4.3 Proof of Cercignani's Conjecture in a Special Case

In fact, with the tools presented before, it is possible to prove the following theorem which bridges a gap (or rather fills a hole) between Theorems 2 and 3.

Theorem 3-ε. *Assume $B(v - v_*, \sigma) \geq K_B(1 + |v - v_*|^2)$, and let f be a probability density on $\mathbb{R}^N$ with $\int f \begin{pmatrix} 1 \\ v \\ |v|^2 \end{pmatrix} dv = \begin{pmatrix} 1 \\ 0 \\ N \end{pmatrix}$. Then,*

$$D(f) \geq \left(K_B \frac{|S^{N-1}|}{8} \frac{N-1}{1+2N}\right) T_f H(f \mid M),$$

where $T_f = \inf\limits_{e \in S^{N-1}} \int_{\mathbb{R}^N} f(v)(v \cdot e)^2 dv$.

Remark 1. This is the first case in which Cercignani's conjecture holds true! (It was discovered during the process of preparing this course, in an attempt to answer some questions by Bodineau.)

Remark 2. Besides its own interest, this result is the starting point for the simplified proof of Theorem 2 which was mentioned above and which is written in [87].

Proof. As before, let us introduce the Fokker–Planck regularization semi-group and start with

$$D(f) \geq \frac{|S^{N-1}|}{4} \overline{D}(f) \geq \frac{K_B |S^{N-1}|}{4} \overline{D}(f) \quad (\overline{D} \text{ with kernel } (1 + |v - v_*|^2))$$

$$-\frac{d}{dt} \overline{D}(S_t f) = \int_{\mathbb{R}^{2N}} \psi(X) \left| \frac{\nabla_x S_t f}{S_t F} - \frac{\nabla_x S_t G}{S_t G} \right|^2 (S_t F + S_t G) \, dX$$

$$- \int_{\mathbb{R}^{2N}} (L^* \psi)(S_t F - S_t G) \log \frac{S_t F}{S_t G} dx,$$

$$F = ff_*, \quad G = \oint f' f'_* d\sigma$$

$$X = (v, v_*), \quad \psi(X) = 1 + |v - v_*|^2, \quad L^* = \Delta_X - X \cdot \nabla_X.$$
$$L^* \psi = 4N - 2|v - v_*|^2 \leq 4N\psi.$$

Recall that

$$\int |v - v_*|^2 \left| \frac{\nabla_X S_t F}{S_t F} - \frac{\nabla_X S_t G}{S_t G} \right|^2 S_t F \, dX \geq \left(\frac{N-1}{2}\right) T_{S_t f} I(S_t f \mid M)$$

$$\geq \frac{N-1}{2} T_f I(S_t f \mid M)$$

$$-\frac{d}{dt}\overline{D}(S_t f) \geq \frac{N-1}{2} T_f I(S_t f \mid M) - 4N\overline{D}(S_t f).$$

From the Blachman–Stam inequality [using $S_t f = M_{1-e^{2t}} * f_{e^{2t}}$ with the notation $g_\lambda = \frac{1}{\lambda^{N/2}} g(\frac{\cdot}{\sqrt{\lambda}})$] or elementary Γ_2 theory,

$$I(S_\tau g \mid M) \leq e^{2\tau} I(g \mid M).$$

In particular, $I(S_{t+2Nt} f \mid M) \leq e^{-4Nt} I(S_t f \mid M)$. Thus

$$-\frac{d}{dt}\left[e^{-4Nt}\overline{D}(S_t f)\right] \geq \frac{N-1}{2} T_f \, I(S_{(1+2N)t} f \mid M).$$

Now take $\displaystyle\int_0^{+\infty} dt$, to find

$$
\begin{aligned}
\overline{D}(f) = \overline{D}(S_0 f) \; &\geq \; \frac{N-1}{2} T_f \int_0^{+\infty} I(S_{(1+2N)t} f \mid M) dt \\
&= \frac{N-1}{2} T_f \int_0^{+\infty} I(S_t f \mid M)\frac{dt}{1+2N} \quad \text{[change of var.'s]} \\
&= \frac{N-1}{2(1+2N)} T_f \, H(f \mid M). \quad \square
\end{aligned}
$$

1.5 The State of the Art for the Spatially Homogeneous Boltzmann Equation

In this chapter, I shall describe how the results of the last chapter can be used to study the trend to equilibrium for the BE, under the very restrictive assumption that f_t does not depend on x. To this end, I shall appeal to many results from the theory of the Cauchy problem for the Boltzmann equation. Explaining precisely how one obtains these results would be a complete course on its own, so I shall not go into this.

1.5.1 Preparations

The theory of the Cauchy problem for the spatially homogeneous BE is complicated and many situations have to be studied separately. More precisely, the theory naturally splits *twice* into two branches:

- Depending on the *kinetic* behavior of the collision kernel $B(v - v_*, \sigma)$ as $|v - v_*| \to 0$ [kinetic here means: as a function of $|v - v_*|$]: one says that the context is of

- Hard potential type if $B \sim |v - v_*|^\gamma \quad \gamma > 0$
- Soft potential type if $B \sim |v - v_*|^\gamma \quad \gamma < 0$
- Maxwellian type if B does not depend on $|v - v_*|$, but only on

$$\cos \theta = \left\langle \frac{v - v_*}{|v - v_*|}, \sigma \right\rangle.$$

The behavior of B as $|v - v_*| \to \infty$ has a decisive impact on the *tail behavior* of solutions: for hard potentials the solutions have a tendency to decrease very fast at infinity, while for soft potential this is not necessarily the case.

- Depending on the *angular* behavior of the collision kernel as $\theta \to 0$ [again, $\theta =$ deviation angle]: one speaks of

  - Cutoff kernel if $\displaystyle \int B(v - v_*, \sigma) d\sigma < +\infty$.

  - Non-cutoff kernel if $\displaystyle \int B(v - v_*, \sigma) d\sigma = +\infty$ (for almost all $|v - v_*|$).

Non-cutoff kernels naturally arise in the modelling of *long-range interactions* (even interactions which would decay extremely fast at large distances). As a general rule, from the stochastic point of view, the cutoff Boltzmann equation represents a *jump process* and is associated with *propagation* of regularity and of singularity; while the non-cutoff one is intermediate between a jump and a *diffusion* process (I refrain from using the words "fractional diffusion", because these terms are commonly used in a completely different meaning), and is associated with regularizing effects. To quantify these statements, more precise assumptions have to be imposed on the kernel.

There may be a third dichotomy in the theory, depending on the behavior of the kernel at *small* relative velocities (singular or not). Nobody is sure...

The theory of cutoff Boltzmann equation is by now rather well-understood; the theory of non-cutoff BE is still under construction (Alexandre, Desvillettes, Fournier, Méléard, Villani, etc.). In these lectures, for simplicity I only consider the cutoff case, in the model situations

$$B(v - v_*, \sigma) = |v - v_*|^\gamma \quad (\gamma > 0),$$

$$B(v - v_*, \sigma) = (1 + |v - v_*|)^{-\beta} \quad (\beta \geq 0).$$

The most important case of application is certainly that of *hard spheres*: in dimension 3, $B(v - v_*, \sigma) = |v - v_*|$ (up to a constant).

1.5.2 Current State of Regularity Theory

By gathering many results from different authors, one obtains

Theorem 5. *Consider the kernel* $B = |v - v_*|^\gamma$, $0 < \gamma \leq 1$, *and let* f *be a probability density on* $\mathbb{R}^N$,
$$\int f(v) \begin{pmatrix} 1 \\ v \\ |v|^2 \end{pmatrix} dv = \begin{pmatrix} 1 \\ 0 \\ N \end{pmatrix}. \text{ Then}$$

(i) *[61] There exists a unique solution of the SHBE* $\partial f/\partial t = Q(f,f)$ *(in the class of solutions whose kinetic energy is nonincreasing – this restriction is important!).*

Moreover $\forall t$
$$\int f(v) \begin{pmatrix} 1 \\ v \\ |v|^2 \end{pmatrix} dv = \begin{pmatrix} 1 \\ 0 \\ N \end{pmatrix}.$$

Remark. Carleman [16] obtained existence and uniqueness under the assumptions f_0 *continuous, radially symmetric,* $f_0(v) = O(1/1 + |v|^6)$, $\gamma = 1$.

(ii) *[37, 61, 92, 95]*
$$\forall s \geq 2 \quad \forall t_0 > 0 \quad \sup_{t \geq t_0} \left\| f_t \right\|_{L^1_s} < +\infty.$$

$\longrightarrow$ *all moments are bounded, uniformly in* $t \geq t_0$.

(iii) *[67]* $\forall t_0 > 0 \quad \exists \rho_0 > 0 \quad \exists A_0 > 0$;
$$t \geq t_0 \implies \forall v \in \mathbb{R}^N \quad f_t(v) \geq \rho_0 e^{-A_0|v|^2}.$$

$\longrightarrow$ *the solution satisfies a Maxwellian lower bound, uniform in* $t \geq t_0$.

Remark. Carleman [16] obtained $\rho_0 e^{-A_0|v|^{2+\epsilon}}$.

(iv) *[46, 47]* $f_0 \in L^1_{s_1} \cap L^p_s$ *(s given,* s_1 *large enough)*
$$\implies \sup_{t \geq 0} \left\| f_t \right\|_{L^p_s} < +\infty \text{ – not explicit.}$$
[64] $f_0 \in L^1_3 \cap L^p$ $(1 < p < \infty)$
$$\implies \forall t_0 > 0 \quad \sup_{t \geq t_0} \left\| f_t \right\|_{L^p} < +\infty \text{ – explicit constants}$$
[3] same as Gustafsson with $p = \infty$ *– explicit constants.*

(v) *[64]* $f_0 \in L^1_3 \cap H^k \quad k \geq 0$
$$\implies \forall t_0 > 0 \quad \sup_{t \geq t_0} \left\| f_t \right\|_{H^k} < +\infty.$$

$\longrightarrow$ *Propagation of* H^k *smoothness. Previous partial results had been obtained by Wennberg [93]:* H^1 *smoothness in some cases, nonexplicit constants,* $N = 3$; *and Carlen et al. [27]: for Maxwellian collision kernels, with* f_0 *having all moments finite.*

Remark. There is no regularization, so $f_0 \notin H^k_{\mathrm{loc}} \implies f_t \notin H^k_{\mathrm{loc}}$. *However, one can prove the following:*

(vi) [64] $f_0 \in L_3^1 \cap L^p$ $1 < p < \infty$ then $\forall k_0 \geq 0$ $\forall s_0 > 0$, $\exists \rho_0, A_0, \lambda_0 > 0$ s.t.

$$t \geq t_0 \Longrightarrow f(t,.) = f^S(t,.) + f^R(t,.)$$
$$\text{"smooth" \quad "remainder"}$$

with

$$\begin{cases} \sup_{t \geq t_0} \left\| f_t^S \right\|_{H^{k_0}} < +\infty, \quad \sup_{t \geq t_0} \left\| f_t^S \right\|_{L_{s_0}^1} < +\infty, \quad f_t^S(v) \geq \rho_0 e^{-A_0|v|^2} \\ \left\| f_t^R \right\|_{L^p} = O(e^{-\lambda_0 t}) \end{cases}$$

The proofs use:

- Moment inequalities (Povzner inequalities...)
- Mixing properties of Q^+ operator, similar to Lions' theorem: under some (very stringent) assumptions on B,

$$\left\| Q^+(g, f) \right\|_{H^{\frac{N-1}{2}}} \leq C \left\| g \right\|_{L^1} \left\| f \right\|_{L^2}$$

- Duhamel's formula (and its iterated versions),

$$f_t(v) = f_0(v) e^{\int_0^t L f_s(v) ds} + \int_0^t Q^+(f_s, f_s)(v) e^{-\int_s^t L f_\tau(v) d\tau} ds$$

$$L f = \left(\int B \, d\sigma \right) * f$$

- A lot of cutting into pieces
- A lot of interpolation in weighted Lebesgue and Sobolev spaces.

1.5.3 Convergence to Equilibrium

Here is an application of the previous regularity results, when combined with the entropy production estimates of the last chapter.

Theorem 6 (Based on Mouhot, Villani). *Let $f_0 \in L_3^1 \cap L^p$ ($1 < p < \infty$),*

$$\int f_0(v) \begin{pmatrix} 1 \\ v \\ |v|^2 \end{pmatrix} dv = \begin{pmatrix} 1 \\ 0 \\ N \end{pmatrix}, \quad B(v - v_*, \sigma) = |v - v_*|^\gamma, \text{ then}$$

$$\left\| f_t - M \right\|_{L^1} = O(t^{-\infty}) \quad \text{with explicit constants}$$

(the statement means: $\forall \epsilon > 0$, $\exists C_\epsilon$ explicit s.t. $\left\| f_t - M \right\|_{L^1} \leq C_\epsilon t^{-1/\epsilon}$).

Proof. Use the decomposition $f_t = f_t^S + f_t^R$ (k_0 to be chosen later on, $t_0 = 1$).
Let t_1 be an intermediate time, to be chosen later; introduce

$$\begin{cases} \widetilde{f}_t = \text{sol. of BE starting from} f_{t_1}^S \text{ at } t = t_1 \\ \widetilde{M} = \text{Maxwellian distribution associated with } \widetilde{f}_{t_1} = f_{t_1}^S. \end{cases}$$

Since $\widetilde{f}_{t_1}$ is very smooth, so is $\widetilde{f}_t$, and in fact $\widetilde{f}_t$ is uniformly bounded in $L_{s_0}^1 \cap H^{k_0}$ for $t \geq t_1 + 1$, with a Maxwellian lower bound. This implies (Theorem 4)

$$D(\widetilde{f}_t) \geq K_\epsilon H(\widetilde{f}_t \mid \widetilde{M})^{1+\epsilon}$$

with ϵ as small as one wishes if k_0 is chosen large enough, depending on ϵ.

As a consequence (Gronwall), $H(\widetilde{f}_t \mid \widetilde{M}) = O((t - t_1)^{-1/\epsilon})$ with explicit constants which *do not depend on* t_1 (they only depend on t_0 used in the decomposition $f^S + f^R$).

On the other hand, from the statement in Theorem 5 we have

$$\left\| f_t^R \right\|_{L^1} = O(e^{-\lambda_0 t}) \quad (t \geq 1)$$

$$\left\| M - \widetilde{M} \right\|_{L^1} = O(e^{-\lambda t_1}) \quad \text{for some } \lambda > 0$$

Next, by a stability result about solutions of the SHBE,

$$\left\| f_t - \widetilde{f}_t \right\|_{L^1} \leq C \, e^{A(t-t_1)} \left\| f_{t_s} - \widetilde{f}_{t_1} \right\|_{L_s^q}$$

[at most exponential divergence]

After a little bit of interpolation, one gets the bound

$$\left\| f_t - M \right\|_{L^1} \leq \left\| f_t - \widetilde{f}_t \right\|_{L^1} + \left\| \widetilde{f}_t - \widetilde{M} \right\|_{L^1} + \left\| \widetilde{M} - M \right\|_{L^1}$$
$$\leq C \left(e^{A(t-t_1)} e^{-\lambda t_1} + (t - t_1)^{-1/\epsilon} + e^{-\lambda t_1} \right).$$

Now, if one chooses $t_1 \gg t - t_1$, the exponential divergence can be compensated for. More precisely, if $A(t - t_1) = \frac{\lambda}{2} t_1$, i.e. $t_1 = \left(\frac{A}{\frac{\lambda}{2} + A} \right) t$, then one gets $\left\| f_t - M \right\| \leq C(t_1) t^{-1/\epsilon}$, as announced. $\square$

Let me conclude this chapter with a result about "mollified soft potentials": $B(v - v_*, \sigma) = (1 + |v - v_*|^2)^{-\beta/2}$ ($\beta \geq 0$), or $B = (1 + |v - v_*|)^{-\beta}$, which gives the same bounds.

Theorem 7 (Toscani and Villani, 1999). *Assume* $B(v - v_*, \sigma) = (1 + |v - v_*|)^{-\beta}$, $0 \leq \beta < 2$. *Let* f_0 *be a probability density satisfying*

$$\int f_0 \begin{pmatrix} 1 \\ v \\ |v|^2 \end{pmatrix} dv = \begin{pmatrix} 1 \\ 0 \\ N \end{pmatrix}, \text{ with } f_0 \in \cap_{s>0} L_s^1, \; f_0 \in L^p \text{ for some } p > 1,$$

$f_0 \geq \rho_0 e^{-A_0|v|^2}$ for some $\rho_0 > 0$, $A_0 > 0$. Then $H(f_t \mid M) = O(t^{-\infty})$.

Remark 1. This theorem is based on Theorem 3 together with the propagation of bounds. These bounds are *not* uniform in time: we only know $\left\| f_t \right\|_{L_s^1} = O((1+t)^\epsilon)$, $\left\| f_t \right\|_{L^p} = O((1+t)^{1+\epsilon})$, and $f_t(v) \geq \rho_t e^{-A_0|v|^2}$ for $\rho_t = e^{Ct}\rho_0$. It turns out that this deterioration of the bounds is *slow enough* to prove the result. This of course is possible only because we have a precise control of the bounds in the entropy production inequalities.

Remark 2. Up to very recently, this was the only known result for trend to equilibrium for collision kernels with that kind of decay – with the exception of Caflisch [14] in a close-to-equilibrium setting. In particular, compactness arguments spectacularly fail in this context because of the non-uniformity in the moment bounds. Very recently, Guo was able to construct close-to-equilibrium solutions for which the moment bounds are uniform, and then both the compactness and the constructive methods can be applied.

1.6 A Review of Some Closely Related Topics

1.6.1 Particles Systems and Kac's Problem

Assume you have an n-particle system governed by microscopic equations, which is well-described in some sense by a macroscopic system as $n \to \infty$. Typically, the macroscopic equations would be those satisfied by the one-particle marginal as $n \to \infty$. If one is interested in the large-time behavior of the particle system, it is natural to (a) carefully investigate the large-time behavior of the macroscopic system, (b) carefully investigate how precise the approximation is in the limit $n \to \infty$ (say, for instance, how close the microscopic system is to a chaotic system governed by the macroscopic equation).

However, in some situations it is possible to directly study the asymptotic behavior of the particle system as $t \to \infty$, uniformly in n. When this is the case, then the large-time behavior of the macroscopic equation can be *deduced* from that of the particle system, and one expects that the limit $n \to \infty$ and $t \to \infty$ are exchangeable.

When Kac first began to think about the rate of convergence for the BE, he thought that a direct approach of the large-time behavior of a particle system might succeed – and would be more satisfactory than the strategy described in the first paragraph. Since the problem for the full Boltzmann equation was much too complicated, he devised simpler models, in which only velocity was taken into account. They are often called "master equations".

Below are two such models. They take the form of evolution equations for the density $f^{(n)}$ of n particles (joint density).

- *The Boltzmann master equation.* The phase space is

$$\{(v_1, \cdots, v_n) \in (\mathbb{R}^N)^n; \ \Sigma|v_i|^2 = 2nE; \ \Sigma v_i = nV\}$$
$$(E = \text{ mean kinetic energy}; \ V = \text{ mean velocity})$$

and the equation is

$$\frac{\partial f^{(n)}}{\partial t} = n \sum_{ij}' \fint_{S^{n-1}} d\sigma \, B(v_i - v_j, \sigma) \left[A_\sigma^{ij} f^{(n)} - f^{(n)} \right]$$

$$\left[A_\sigma^{ij} f^{(n)} \right] (v_1, \ldots, v_n) = f^{(n)}(v_1, \cdots, v_i', \cdots, v_j', \ldots, v_n)$$

$$\begin{cases} v_i' = \dfrac{v_i + v_j}{2} + \dfrac{|v_i - v_j|}{2} \sigma, \\ v_j' = \dfrac{v_i + v_j}{2} - \dfrac{|v_i - v_j|}{2} \sigma, \end{cases}$$

where $\fint_{S^{N-1}} d\sigma = \dfrac{1}{|S^{N-1}|} \int d\sigma$, $\displaystyle\sum_{ij}' = \dfrac{1}{\binom{n}{2}} \sum_{i<j}$ [If the initial datum

$f_0^{(n)}$ is chaotic.]
In the limit $n \to \infty$, it is expected that the 1-particle marginal $f^{(1)}$ of $f^{(n)}$ converges towards a solution of the Boltzmann equation

$$\frac{\partial f}{\partial t} = \fint_{S^{N-1}} d\sigma \int dv_* B(v - v_*, \sigma)(f'f_*' - ff_*).$$

Remark. In this model v_i', v_j' should be thought of as the velocities *after* collision, instead of *before* for the Boltzmann equation.
The limit $n \to \infty$ has been studied by various authors, including Kac himself, Sznitman, Méléard and collaborators.

- *The Kac master equation.* This is a simplified model of the above. The phase space is $\{(v_1, \cdots, v_n) \in \mathbb{R}^N; \ \Sigma v_i^2 = 2nE\} = \sqrt{2nE}\, S^{n-1}$, and the equation is

$$\frac{\partial f^{(n)}}{\partial t} = n \sum_{ij}' \fint_0^{2\pi} \left[f_0^{(n)} \circ R_\theta^{ij} - f^{(n)} \right] d\theta, \tag{K}$$

where $R_\theta^{ij}(v_1, \ldots, v_n) = (v_1, \ldots, v_i', \ldots, v_j', \ldots, v_m)$ and $(v_i', v_j') = R_\theta(v_i, v_j)$, where R_θ is the direct rotation with angle θ in the plan $\mathbb{R}^2$.
For notational simplicity let us set $E = \dfrac{1}{2}$, so the phase space is $\sqrt{n} S^{n-1}$.
Let us focus on Kac's master equation; the discussion would be similar for the Boltzmann master equation. If one rewrites (K) in the form

$$\frac{\partial f^{(n)}}{\partial t} = L_n f^{(n)},$$

then the operator L_n is a beautiful linear operator, symmetric in $L^2(d\sigma^n_{\sqrt{n}})$ where $\sigma^n_{\sqrt{n}}$ is the uniform probability measure on $\sqrt{n}S^{n-1}$. Therefore Kac thought is was possible to attack the problem of trend to equilibrium by the apparatus of linear operator theory. It is easy to see that for each n, L_n is a nonpositive operator, admitting 0 as a simple eigenvalue, associated to the eigenfunction 1 (conservation of mass). Moreover, 0 is an isolated eigenvalue, so one can define

$$\lambda_n = \text{the gap between 0 and the largest nonzero eigenvalue.}$$

This naturally suggests
Kac's spectral gap problem. Find a uniform lower bound on λ_n.
Here uniform means of course uniform as $n \to \infty$. This problem stood open from 1956 to 1999, the date when it was solved by Janvresse, by the use of Yau's martingale method (see [50]). Very shortly after, the result was spectacularly improved by Carlen et al. [22]:

Theorem 8 (Carlen, Carvalho, Loss). $\Lambda_n = \dfrac{n+2}{2(n-1)}$ *Moreover the corresponding eigenfunction is* $\displaystyle\sum_{j=1}^{n} v_j^4 - \dfrac{3}{n(n+2)}.$

A more precise description of the spectrum was actually obtained before, by a very different method, in a preprint by Maslen which remained unpublished for a long time. Recently, Carlen, Carvalho and Loss obtained a result rather similar to Theorem 8 for the Boltzmann master equation, even though less explicit (in the case where $B(v - v_*, \sigma)$ only depends on $\langle v - v_*/|v - v_*|, \sigma\rangle$). Their method is very interesting and certainly deserves further study; it was applied recently by Caputo (2003) to solve a tricky spectral gap problem for spin systems.

However, once the result is proven, it is not clear at all what can be deduced from it for the nonlinear Boltzmann equation, or for its variant for Kac's master equation, which is the Kac model

$$\begin{cases} \dfrac{\partial f}{\partial t} = \displaystyle\int_{\mathbb{R}} \int_0^{2\pi} (f'f'_* - ff_*)\, d\theta\, dv_*, \\ (v', v'_*) = R_\theta(v, v_*). \end{cases} \qquad (\text{Km})$$

The problem is that, for instance, one has

$$\left\| f^{(n)} - 1 \right\|_{L^2(d\sigma^n_{\sqrt{n}})} \leq e^{-\lambda t}$$

for some $\lambda > 0$ independent of n. But in the limit going to the BE, the L^2 norm blows up very quickly, something like C^n for some constant C. So one cannot a priori conclude that some quantity, converging as $n \to \infty$ towards something, will decrease exponentially fast. Certainly the problem lies in the limit which is not adapted to L^2 norm – note that $L^2(d\sigma^n_{\sqrt{n}})$ transform into $L^2(M^{-1}dv)$ at the level of $f^{(1)}$. I personally think that the right thing to do here is to go to the *linearized* Boltzmann equation by a different scaling.

Remark. As a corollary of his study, Kac was led to the following strange conjecture, which I have never seen mentioned anywhere except in his text. Whenever f solves (Km), or maybe the BE with Maxwell molecules, and has zero mean velocity and unit temperature, then the following functional $K(f)$ would be nonincreasing with time:

$$
\begin{cases}
K(f) = \displaystyle\int e^{z_0(f^2)(1-v^2)} f^2(v) \, dv, \\[2mm]
z_0(f^2) \text{ sol. of } \displaystyle\int (1 - v^2) e^{-z_0(f^2)v^2} f^2(v) \, dv = 0.
\end{cases}
$$

To investigate rates of convergence towards the BE, a better idea is to look for the entropy production versions of Kac's problem.

Entropy variant of Kac's spectral gap problem. Let μ_n be the best constant in the inequality

$$
D(f^{(n)}) \geq \mu_n \left[H(f^{(n)}) - H(1) \right] = \mu_n H(f^{(n)}).
$$

Is there a uniform lower bound on μ_n?

Here $D(f^{(n)}) = \displaystyle\sum_{ij}' \frac{n}{2} \int_{\sqrt{n}} d\theta \int_{S^{n-1}} \left(f^{(n)} \circ R^{ij}_\theta - f^{(n)} \right) \log \frac{f^{(n)} \circ R^{ij}_\theta}{f^{(n)}} d\sigma^n_{\sqrt{n}}(v);$

of course this is the entropy production functional associated to Kac's master equation.

If there is a uniform lower bound $\mu > 0$ for μ_n, then

$$
H\left(f_t^{(n)} \right) \leq H\left(f_0^{(n)} \right) e^{-\mu t}.
$$

If there is a strong enough chaos property, then $f_t^{(n)} d\sigma^n$ looks like $[f_t(v)dv]^{\otimes n}$, where f_t is a solution of the Boltzmann equation. A well-known lemma states that $d\sigma^n_{\sqrt{n}}$ roughly speaking, looks like $M^{\otimes n}$, where M is the standard Gaussian distribution on $\mathbb{R}$. Then

$$
H\left(f_t^{(n)} \right) = \int f_t^{(n)} \log f_t^{(n)} d\sigma^{(n)}_{\sqrt{n}} \simeq \int \frac{f_t^{\otimes n}}{M^{\otimes n}} \log \frac{f_t^{\otimes n}}{M^{\otimes n}} M^{\otimes n} dv?
$$

$$
\simeq nH\left(f_t \mid M \right).
$$

(Here I just encountered notational problems because $f_t^{(n)}$ is a density with respect to $d\sigma^n$ which is morally a Gaussian measure, while f_t is a density with respect to Lebesgue measure, etc.)

Then, providing everything passes to the limit we would have

$$H\Big(f_t \mid M\Big) \le H\Big(f_0 \mid M\Big)e^{-\mu t},$$

as the limit of

$$\frac{H\Big(f_t^{(n)}\Big)}{n} \le \frac{H\Big(f_0^{(n)}\Big)}{n}e^{-\mu t}.$$

However, this would probably be asking too much: passing to the limit in the entropy–entropy production inequality for fixed n, we would recover an entropy–entropy production inequality of linear type for Kac's model, which would contradict the intuition furnished by the Bobylev–Cercignani counterexamples. Actually, Carlen, Carvalho and Loss recently checked that the optimal μ_n goes to 0 as $n \to \infty$. However, the picture is still not so clear because the Bobylev–Cercignani counterexamples exploit quite well the structure of the Boltzmann operator, and cannot be directly adapted to the context of the Kac equation. In fact, after working on that question with Carlen, Carvalho and Loss I have come to suspect that linear entropy–entropy production inequalities may after all exist for the Kac equation under some moment conditions; there is something to understand here.

Anyway, here is now some positive result: it was shown in my Habilitation (2000), that $\mu_n^{-1} = O(n)$.

So one can ask:

- What is the exact order of μ_n?
- Whether there are some initial data $f_0^{(n)}$ which provide a better rate than μ_n.
- Whether there are related models with μ_n bounded below.

Here is an answer to the third question. The next theorem was proven for the sake of this course, then slightly generalized in [87].

Theorem 9. *Consider the following modified Kac master equation:*

$$\frac{\partial f^{(n)}}{\partial t} = n \sum_{ij}' \int_0^{2\pi} (1 + v_i^2 + v_j^2)[f^{(n)} \circ R_\theta^{ij} - f^{(n)}]$$

[Note that v_i^2, v_j^2 are typically $O(1)$ if $v \in \sqrt{n}S^{n-1}$, so this equation has the same physical scaling as the previous one.]

Define $H\left(f^{(n)}\right) = \displaystyle\int_{\sqrt{n}S^{n-1}} f^{(n)} \log f^{(n)} \, d\sigma_{\sqrt{n}}^n \quad \left(\displaystyle\int f^{(n)} d\sigma^n = 1\right)$

$$D\left(f^{(n)}\right) = \frac{n}{2} \int_{\sqrt{n}S^{n-1}} d\sigma^n_{\sqrt{n}} \sum_{ij}' \int_0^{2\pi} d\theta (1 + v_i^2 + v_j^2) \left[f^{(n)} \circ R_\theta^{ij} - f^{(n)}\right] \log \frac{f^{(n)} \circ R_\theta^{ij}}{f^{(n)}}$$

Then $D(f^{(n)}) \geq \dfrac{n}{5n-1} H(f^{(n)}) \geq \dfrac{1}{4} H(f^{(n)})$.

Proof. To simplify notations let us rescale everything to work on the unit sphere S^{n-1}. Let σ^n be the uniform probability measure on S^{n-1}. The problem becomes: bounding below

$$D\left(f^{(n)}\right) = \frac{n}{2\binom{n}{2}} \sum_{i<j} \frac{1}{2\pi} \int_0^{2\pi} d\theta \int_{S^{n-1}} (1 + nv_i^2 + nv_j^2) \left[f^{(n)} \circ R_\theta^{ij} - f^{(n)}\right] \log \frac{f^{(n)} \circ R_\theta^{ij}}{f^{(n)}}$$

in terms of $H(f^{(n)}) = \displaystyle\int_{S^{n-1}} f^{(n)} \log f^{(n)} \, d\sigma^n$, whenever $\displaystyle\int f^{(n)} d\sigma^n = 1$.

The proof follows the same lines as Theorem 3-ϵ. First, by Jensen,

$$D(f^{(n)}) \geq \overline{D}(f^{(n)}) \equiv \frac{n}{2\binom{n}{2}} \sum_{i<j} \int_{S^{n-1}} (1 + nv_i^2 + nv_j^2)(f - f^{ij}) \log \frac{f}{f^{ij}} \, d\sigma^n$$

where $f^{ij}(v) = \dfrac{1}{2\pi} \displaystyle\int_0^{2\pi} f(R_\theta^{ij} v) d\theta$.

Let Δ_S be the Laplace–Beltrami operator on the sphere S^{n-1}; by definition (this is one of the possible definitions),

$$\Delta_S f = \sum_{i<j} (D^{ij})^2 f,$$

where $D^{ij} f = v_j \partial_i f - v_i \partial_j f = \dfrac{d}{d\theta}\Big|_{\theta=0} f \circ R_\theta^{ij}$.

An easy computation shows that

$$\int_{S^{n-1}} f \Delta_S f = -\int_{S^{n-1}} \sum_{i<j} (D^{ij} f)^2$$

$$= -\int_{S^{n-1}} |v|^2 |\nabla f|^2 + \int_{S^{n-1}} (v \cdot \nabla f)^2 = -\int_{S^{n-1}} |\nabla f|^2$$

(tangential gradient or gradient on S^{n-1}).

Let $(S_t)_{t \geq 0}$ be a semigroup associated with the heat equation

$$\partial_t f = \Delta_S f.$$

This is the equivalent of the Ornstein–Uhlenbeck semigroup $\partial_t h = \Delta h - v \cdot \nabla h$. A computation similar to that in the proof of Theorem 3-ϵ yields

$$-\frac{\mathrm{d}}{\mathrm{d}t}\overline{D}(S_t f^{(n)})$$

$$= \frac{n}{2\binom{n}{2}} \sum_{i<j} \int_{S^{n-1}} (1 + n v_i^2 + n v_j^2)(S_t f + S_t f^{ij}) |\nabla \log S_t f - \nabla \log S_t f^{ij}|^2 \mathrm{d}\sigma^n$$

$$-\frac{n}{2\binom{n}{2}} \sum_{i<j} \int_{S^{n-1}} \Delta_S (1 + n v_i^2 + n v_j^2)(S_t f - S_t f^{ij}) \log \frac{S_t f}{S_t f^{ij}}$$

Here I used $(S_t f)^{ij} = S_t(f^{ij})$.

Next, $\Delta_S(1 + n v_i^2 + n v_j^2) = n[4 - 2n(v_i^2 + v_j^2)] \le 4n$. And Also, if P_{ij} denotes the orthogonal projection on the (i, j) plane,

$$(S_t f + S_t f^{ij}) \Big| \nabla \log(S_t f) - \nabla \log(S_t f)^{ij} \Big|^2$$

$$\ge S_t f \Big| P_{ij} \nabla \log S_t f - P_{ij} \nabla \log(S_t f)^{ij} \Big|^2$$

But $P_{ij} \nabla \log S_t f^{ij}$ is always colinear to $(v_i, v_j) \in$ plane (i, j).

If T is the operator with unit norm $(\mathbb{R}^2 \to \mathbb{R})$ $[x_i, x_j] \xrightarrow{T} \frac{x_i v_j - x_j v_i}{v_i^2 + v_j^2}$ then $T P_{ij} \nabla \log(S_t f)^{ij} = 0$ and we have

$$|P_{ij} \nabla \log S_t f - P_{ij} \nabla \log(S_t f)^{ij}|^2 \ge |T P_{ij} \nabla \log S_t f|^2$$

$$= \frac{(D^{ij} \log S_t f)^2}{v_i^2 + v_j^2}.$$

All in all,

$$-\frac{\mathrm{d}}{\mathrm{d}t}\overline{D}\Big(S_t f^{(n)}\Big) \ge \frac{n}{2\binom{n}{2}} \sum_{i<j} \int_{S^{n-1}} n S_t f \Big(D^{ij} \log S_t f \Big)^2 \mathrm{d}\sigma^n$$

$$-\frac{n}{2\binom{n}{2}} \sum_{i<j} \int_{S^{n-1}} 4n(S_t f - S_t f^{ij}) \log \frac{S_t f}{S_t f^{ij}} \mathrm{d}\sigma$$

$$= \frac{n^2}{2\binom{n}{2}} \int_{S^{n-1}} S_t f |\nabla \log S_t f|^2 \mathrm{d}\sigma^n - 4n \overline{D}\Big(S_t f^{(n)}\Big)$$

which is

$$-\frac{d}{dt}\overline{D}\left(S_t f^{(n)}\right) + 4n\overline{D}\left(S_t f^{(n)}\right) \geq \frac{n}{n-1} \int_{S^{n-1}} S_t f |\nabla \log S_t f|^2 d\sigma^n$$

$$-\frac{d}{dt}\left[e^{-4nt}\overline{D}(S_t f^{(n)})\right] \geq \frac{n}{n-1} e^{-4nt} \int_{S^{n-1}} S_t f |\nabla \log S_t f|^2 d\sigma^n$$

$$= \frac{n}{n-1} e^{-4nt} I\left(S_t f^{(n)}\right)$$

Here $I(f) = \int \frac{|\nabla f|^2}{f} d\sigma^n$. From the theory of logarithmic Sobolev inequalities, $I(S_t f) \leq e^{-\rho t} I(f)$ with $\rho = \frac{n-1}{(\text{radius})^2} = n-1$. So

$$I\left(S_{t+\frac{4n}{n-1}t} f\right) \leq e^{-4nt} I(S_t f),$$

and we have

$$-\frac{d}{dt}\left[e^{-4nt}\overline{D}(S_t f^{(n)})\right] \geq \frac{n}{n-1} I\left(S_{(1+\frac{4n}{n-1})t} f^{(n)}\right)$$

If one $\int_0^{+\infty} dt$, one finds

$$\overline{D}\left(f^{(n)}\right) \geq \frac{n}{n-1} \int_0^{+\infty} I\left(S_{(1+\frac{4n}{n-1})t} f^{(n)}\right) dt = \left(\frac{n}{n-1}\right)\frac{n-1}{5n-1} \int_0^{+\infty} I(S_t f)$$

$$= \frac{n}{5n-1} H\left(f^{(n)}\right). \quad \square$$

Physical interpretation. This model enhances the relaxation of *fast* particles, and this is precisely what the Boltzmann equation needs for the fast decrease of entropy.

1.6.2 The Central Limit Theorem for Maxwellian Molecules

In the particular case when $B(v - v_*, \sigma) = b(\cos\theta)$ [$\theta = $ deviation angle as usual], then there is a strong analogy between the Q^+ operator

$$Q^+(f, f) = \int dv_* \, d\sigma \, b(\cos\theta) f' f'_*$$

and the rescaled convolution operator

$$f_{1/2} * f_{1/2} = \frac{1}{2^{N/2}} f\left(\frac{\cdot}{\sqrt{2}}\right) * \frac{1}{2^{N/2}} f\left(\frac{\cdot}{\sqrt{2}}\right).$$

Thus, many tools which work (more or less classically) for the study of rate of convergence in the central limit theorem, also have a chance to work in the study of the Boltzmann equation.

McKean [60] worked out the brilliant idea, apparently first suggested by Kac, to explicitly formulate the asymptotic behavior for the BE (spatially homogeneous with Maxwellian molecules!!) in a form similar to the CLT. The idea was that f_t could be written as a sum of terms taking into account all possible collisional scenarii. One term would take into account particles who have never collided, another one particles which have collided once with a particle having never collided before, etc. It is easy to have such representation by *iteration of Duhamel's formula*: starting from

$$\frac{\partial f}{\partial t} = Q^+(f, f) - f$$

(f is the simple form taken by the loss term, due to the particular collision kernel and the normalization), one gets

$$f_t(v) = f_0(v)e^{-t} + \int_0^t e^{-(t-s)} Q^+(f_s, f_s)\, ds$$

then one can replace f_s in the integral by $f_0 e^{-s} + \int_0^s e^{-(s-s_2)} Q^+(f_{s_2}, f_{s_2})\, ds_2$.
If one does this an infinite number of times, one arrives at the
   Wild sum representation.

$$f_t = \sum_{n=1}^{\infty} e^{-t}(1 - e^{-t})^{n-1} \sum_{\gamma \in \Gamma(n)} \alpha(\gamma)\, Q_\gamma^+(f_0)$$

- n is the number of particles involved
- $e^{-t}(1 - e^{-t})^{n-1}$ are coefficients which go to 0 as $t \to \infty$, for given n
- γ is a tree with n nodes, describing the collision history for the n particles
- α is a combinatorial coefficient
- everything is expressed in terms of the initial datum f_0

Here $\Gamma(n)$ is the set of all binary trees with n leaves, $Q_\gamma^+(f_0)$ is defined by induction as follows: let $\gamma \in \Gamma(n)$, then γ defines two sub-trees (take the root out), γ_1 and γ_2. Then

$$Q_\gamma^+(f_0) = Q^+(Q_{\gamma_1}^+(f_0), Q_{\gamma_2}^+(f_0)),$$

with the definition $Q^+(g, f) = \int b(\cos \theta) f' g'_* \, dv_* \, d\sigma$. If $n = 0$ one agrees $Q^+_\gamma(f_0) = f_0$. The term f_0 corresponds to particles having undergone no collision (so they are still in their initial state). The term $Q^+(f_0, f_0)$ to particles having undergone one collision; etc. For instance,

$$Q^+_\gamma(f_0) = Q^+(Q^+(f_0, Q^+(f_0, f_0)), \, Q^+(f_0, f_0))$$

stands for particles having undergone, first a collision with another particle having undergone no collision, second a collision with a particle who had previously twice collided a particle having undergone no collision before (Exercise: draw the corresponding tree).

Finally, $\alpha(\gamma)$ is a combinatorial coefficient, recursively defined by

$$\alpha(\gamma) = \frac{\alpha(\gamma_1)\alpha(\gamma_2)}{n - 1}; \qquad \alpha(\gamma) = 1 \text{ if } n = 1.$$

(γ_1, γ_2 are the subtrees of γ).

Intuitively, each time a particle undergoes a collision, this will take it closer to equilibrium. This is not really true: the collision partner should not be too far from equilibrium itself. The worst example is a tree in which each node has a terminal leaf; then $Q^+_\gamma(f_0) = Q^+(f_0, Q^+(f_0, Q^+(f_0, \ldots)))$, which certainly is not close to equilibrium... The "good" scenario is one in which all particles undergo many collisions, so that all branches in the tree are quite long. So the game suggested by McKean is to:

(1) Show that if a tree is well-balanced (all particle undergo many collisions), then $Q^+_\gamma(f_0)$ is close to $M =$ equilibrium.
(2) Show that in the wild sum representation, well-balanced trees will play the most important role as $t \to \infty$.

This looks quite delicate. Nevertheless, McKean (1966) was able to complete this program in the case of the Kac ($N = 1$) model (getting exponential convergence). The key estimate was

$$\sum \left\{ \alpha(\gamma); \sup_{a < b} \left| \int_a^b Q^+_\gamma(f_0) - \int_a^b M \right| > \delta \right\} \le c(\delta, f_0) n^{-\delta}, \qquad \delta = 1 - (8/3\pi).$$

In higher dimension, McKean failed to get similar estimates. The problem was recently solved in a lot of generality by Carlen et al. [21].

How can one quantify the fact that collisions have a tendency to drive the system to equilibrium? The key point is to find some inequality expressing the fact that $Q^+(f, g)$ is closer to equilibrium than f, or g. The previous authors prove that if

$$\Phi(f) = \left\| (f - P_f) - M \right\|_\alpha + K \sqrt{\sum_{ij} p_{ij}^2}$$

$$P_f = \mathcal{F}^{-1}\left(-\frac{1}{2}\sum_{ij} p_{ij}\xi_i\xi_j\psi(|\xi|)\right)$$

$$M \quad \text{Maxwellian}$$

with $p_{ij} = \int_{\mathbb{R}^N}\left[v_iv_j - \frac{1}{N}d_{ij}|v|^2\right]f(v)dv$ and $\mathcal{F}^{-1} =$ inverse Fourier transform and ψ is a smooth, monotone decreasing function, $\psi \equiv 1$ close to 0, $\psi \equiv 0$ at infinity and K is large enough and $\|F\|_\alpha = \sup\dfrac{|\mathcal{F}(F)(\xi)|}{|\xi|^{2+\alpha}}$ ($\mathcal{F} =$Fourier transform), then

$$\exists c \in (0,1); \ \Phi(Q^+(f,g)) \le \frac{c}{2}[\Phi(f) + \Phi(g)]$$

This is closely related to several interesting formulas in the same spirit:

Tanaka's theorem. $W_2(Q^+(f,f), Q^+(g,g)) \le W_2(f,g)$; in particular $W_2(Q^+(f,f), M) \le W(f,M)$. Here W_2 is the (improperly called) Wasserstein distance with exponent 2

$$W_2(f,g) = \sqrt{\inf\int_{\mathbb{R}^N\times\mathbb{R}^N}|v-w|^2\mathrm{d}\pi(v,w)}$$

where the inf runs over all π's with marginal densities f and g.

Nonexpansivity of Q^+ in Toscani's distance. $d_2(Q^+(f,f), Q^+(g,g)) \le d_2(f,g)$ where $d_2(f,g) = \sup\limits_{\xi\in\mathbb{R}^N}\dfrac{|\hat{f}(\xi) - \hat{g}(\xi)|}{|\xi|^2}$ (hat stands for Fourier transform). This can be improved, as shown by Gabetta et al. [44] and by Carlen [27] by looking at

$$d_s(f,g) = \sup\limits_{\xi\in\mathbb{R}^N}\frac{|\hat{f}(\xi) - \hat{g}(\xi)|}{|\xi|^s}$$

for $s > 2$. Since this expression only makes sense if the Taylor polynomials of $\hat{f}$ and $\hat{g}$ at $\xi = 0$ coincide up to order $[s] + 1$, it is necessary to subtract a well-chosen polynomial to f before applying d_s to the solution of the BE.

Spectacular results have been obtained by these Fourier-based metrics: in particular, Carlen et al. [27] show that, if f_0 has all its moments bounded, as well as all its Sobolev norms, then

$$\left\|f_t - M\right\|_{L^1} = O\left(e^{-(\lambda-0)t}\right)$$

($\lambda =$ spectral gap of the linearized Boltzmann operator) which is almost optimal. The strong assumptions on the smoothness and moments are used to interpolate between the distance d_s and smoothness/moments, to recover convergence in L^1 norm. I think that the assumptions on the initial datum may be considerably relaxed by a study similar to that in [64].

Stam–Boltzmann inequalities. These are the analogues of the Stam inequalities:

$$H\Big(Q^+(f,g)\Big) \le \frac{1}{2}\Big[H(f) + H(g)\Big]$$

$$I\Big(Q^+(f,g)\Big) \le \frac{1}{2}\Big[I(f) + I(g)\Big]$$

They were proven for general Maxwell kernels in [80]. Earlier works by Carlen and Carvalho (1992), and Toscani [75, 76] considered particular cases. In particular, in the case of a particular kernel, Carlen and Carvalho gave an interesting *lower bound* on $-\Big[H(Q^+(f,f)) - H(f)\Big]$. This was the starting point of their theory for trend to equilibrium, based on the important remark that if $B(v - v_*, \sigma) = b(\cos\theta)$, $\int b(\cos\theta)\mathrm{d}\sigma = 1$, then

$$D(f) \ge H(f) - H\Big(Q^+(f,f)\Big).$$

(D = entropy production).

As we saw in previous chapters, now there are very good bounds for $D(f)$ – but what for $H(f) - H(Q^+(f,f))$? Recent results by Ball et al. [5] suggest that a bound like

$$H(f) - H(Q^+(f,f)) \ge \mathrm{const.}[H(f) - H(M)]$$

can be hoped for under *very* strong conditions on f, namely that it satisfy a Poincaré inequality:

$$\int fh = 0 \Longrightarrow \int f|\nabla h|^2 \ge \mathrm{const.} \int fh^2.$$

The story here is certainly not finished.

1.6.3 The Role of High Energy Tails

In the particular case of Maxwellian molecules, the role of high energy tails to hinder the convergence to equilibrium has been illustrated by rather spectacular works of Carlen and Lu [28] and Carlen et al. [26]. Carlen and Lu showed how to construct solutions converging *arbitrarily slowly* to equilibrium, say $\|f(t) - M\|_{L^1} \ge K/\log\log\log\log\log t$ as $t \to \infty$; this is achieved by means of very heavy distributional tails at high velocities (still with finite energy). Then Carlen, Gabetta and Regazzini showed that if the initial datum has infinite energy, there is no convergence to equilibrium at all; instead, all the energy goes to high velocities, in a way which can be somewhat quantified.

There exist related works by Bobylev and Cercignani [10–12], constructing "eternal" self-similar solutions (with infinite energy) of the spatially homogeneous Boltzmann equation.

In all these contributions, Maxwellian interactions play a very particular role and some of these results (maybe all of them) do not hold for hard potentials.

1.6.4 Behavior of the Fisher Information

As a consequence of the Stam–Boltzmann inequalities, one can show that

$$\frac{\mathrm{d}}{\mathrm{d}t} I(f_t) \leq 0$$

for Maxwell collision kernels. This was first proven by McKean (1966) for the Kac model, then Toscani [75, 76] and Carlen and Carvalho (1992) for other special cases; finally I treated the general case (1998). A few years ago, this could be considered as a precious regularity estimate; but now much stronger regularity estimates have appeared in the spatially homogeneous case.

There is an interesting connection between this problem and the proof of Theorem 3-ϵ or 3. In the case of Maxwell molecules, the Boltzmann semigroup $(B_t)_{t\geq 0}$ commutes with the Fokker–Planck semigroup $(S_t)_{t\geq 0}$ (this observation goes back to Bobylev). On the other hand,

$$I(S_t f \mid M) = -\frac{\mathrm{d}}{\mathrm{d}t} H(S_t f \mid M).$$

Therefore

$$\frac{\mathrm{d}}{\mathrm{d}s}\bigg|_{s=0} I(B_s f \mid M) = -\frac{\mathrm{d}}{\mathrm{d}s}\bigg|_{s=0} \frac{\mathrm{d}}{\mathrm{d}t}\bigg|_{t=0} H(B_s S_t f \mid M)$$

$$= \frac{\mathrm{d}}{\mathrm{d}t}\bigg|_{t=0} \frac{\mathrm{d}}{\mathrm{d}s}\bigg|_{s=0} H(S_t B_s f \mid M)$$

$$= \frac{\mathrm{d}}{\mathrm{d}t}\bigg|_{t=0} D(S_t f).$$

This surprising relation was noticed by McKean (1966), later rediscovered by Toscani and Villani [79]. It implies that $\frac{\mathrm{d}}{\mathrm{d}t} D(S_t f) \leq 0$ for Maxwell kernel, and confirms the fact that the Fokker–Planck semigroup is a good thing to let act on $D(f)$!

1.6.5 Variations on a Theme

The H Theorem states that the H functional goes down in time along solutions of Boltzmann's equation; and we just saw that, at least in a particular case, the Fisher information I also goes down in time along these solutions. It is natural to ask if there is more to uncover here. There are in fact several possible generalizations, of various interest.

More general equations. The decrease of the Fisher information can be easily extended to the Landau equation with Maxwellian molecules, either directly [83], or by passing to the limit of grazing collisions, which relates the Boltzmann and Landau equation. But one can ask whether the restriction to Maxwellian molecules is mandatory. My personal guess is that the decreasing property is not true for general kernels, although I have been somewhat surprised to see numerical simulations suggesting a decreasing behavior for the Landau equation with Coulomb interaction.

At the level of the linear Fokker–Planck equation, it is remarkable that the (relative) Fisher information is indeed decreasing: this is part of the Γ_2 theory of Bakry and Emery, and was underlying McKean's work in dimension 1. Actually, Toscani [77] showed that a simple adaptation of a trick by McKean yields the result in any dimension. This is somehow the method which I adapted in [83] to a mildly nonlinear setting.

For the linear Fokker–Planck equation, the decrease of the Fisher information means the convexity (in time) of the Kullback information. McKean [60] suggested that for Boltzmann's equation, or at least Kac's caricature, the H functional would be not only decreasing with time, but also a convex and even a completely monotone function of time (all derivatives would have alternate signs). This "super-H-Theorem" immediately attracted the attention of the physicists' community, and some people tried to prove it for various models. There was a kind of thrill when it was discovered that certain particular solutions constructed by Bobylev and others did satisfy such a monotonicity property for derivatives up to order about 100; nevertheless the conjecture was in the end disproved. References on this episode can be easily traced from Lieb [53] and Olaussen [65]. Then physicists withdrew from this problem; to my knowledge, even convexity in time has not been proved or disproved. Numerical simulations often show a curve which vaguely seems convex, in the spatially homogeneous case. On the contrary, there is rather good numerical evidence that such a convexity property is downright false for spatially inhomogeneous equations, and one can also construct explicit examples in the linear case (see Sect. 7 for some intuition).

A final comment about this issue: It so happens that a decade ago, a theoretical physicist named Frieden predicted that Fisher's information would be decreasing for a large variety of physical systems, and in fact developed a very peculiar vision of the world, with Fisher's information as a unifying principle for most of quantum and classical physics, plus other sciences as well. After publishing his book *Physics from Fisher information: a Unification* (Cambridge University Press, 1998), he recently released the second edition as *Science from Fisher information: a Unification* (2004)[1]. There is actually a small community of researchers working actively on the subject (A. Plastino, A.R. Plastino, B.H. Soffer, etc.). I don't have any opinion on the relevance of this line of research, but just conclude that the decreasing property of the Fisher information along certain physical systems was expected by some, independently of McKean's contribution.

More general functionals. Could the H and I functionals be just the first two members of a large family of functionals which are all decreasing in time along solutions of, say, the spatially homogeneous Boltzmann equation with Maxwellian kernel? This was more or less what McKean [60] suggested for the Kac caricature, considering the family of functionals obtained by successively differentiating the H functional along the heat equation: define $I_n(f)$ as $(-d/dt)^n H(e^{t\Delta} f)$, hope that I_n is nonincreasing with time. This apparently crazy suggestion becomes less surprising if one recalls McKean's interest in completely integrable partial differential equations. Nobody has proven or disproven this conjecture, but there is reason to believe it is false after Ledoux's formal study (1995) [52] on the linear Fokker–Planck equation.

Functionals involving two solutions. A very striking remark was made by Plastino and Plastino [66]: if f_t and g_t are two solutions of the linear Fokker–Planck equation, then not only are $H(f_t|M)$ and $H(g_t|M)$ nonincreasing functions of t, but so is also $H(f_t|g_t)$. I am not aware of any important application, but this property is "too striking to be useless". It would be interesting to know whether this "nonexpansivity" property, which is somewhat reminiscent of Tanaka's work, also holds true for, say, the Boltzmann equation with Maxwellian molecules.

1.7 Wiping out Spatial Inhomogeneities

If one is interested in estimating rates of convergence in the hydrodynamic limit, one should only retain from the preceding discussion that inequalities like

$$D(f) \geq K_f H(f \mid M^f)^\alpha$$

are a quantitative way to express statements like: small entropy production $\Longrightarrow f \simeq M^f$.

This is a *local* statement, "local" meaning "for each x". In the hydrodynamic limit, one only wants to control the local behavior of f and establish the asymptotic equation satisfied by M^f, or equivalently by ρ, u, T.

---

[1] Some more detailed information on these topics can be found on Frieden's Web page at `http://www.optics.arizona.edu/faculty/resumes/frieden.htm`. Not all readers did appreciate the book, as can be seen from the severe review at `http://cscs.umich.edu/~crshalizi/reviews/physics-from-fisher-info/`, but many did find it valuable, and some consider it as a milestone in theoretical physics.

On the other hand, if one is interested in the trend to equilibrium problem, in a spatially-dependent context, then one wants to establish that *globally* f looks like a Maxwellian as $t \to \infty$; i.e., there exists a Maxwellian distribution $M(v)$, independent on x, such that $F_t \xrightarrow[t \to \infty]{} M$.

Entropy production estimates can only give an information on how close f is from local equilibrium, i.e. $f \simeq M^f = M(\rho, u, T))$, for some $\rho(x), u(x), T(x)$. In fact, if f is any $M(\rho, u, T)$, then $D(f) = 0$. So if one wants to show $f_t \to M$, then one has to introduce additional input.

The theory of the Cauchy problem for the full Boltzmann equation is *incredibly* difficult, and situations where one can prove existence of smooth, well-behaved solutions are extremely rare. However, it seems reasonable (to me) to conjecture the following:

Conjecture 1. Let $B(v - v_*, \sigma) = |v - v_*|^\gamma b(\cos \theta)$ (say) with $\gamma > 0$, b integrable, and let f_0 be a smooth (C^∞, rapidly decaying) nonnegative density on $\Omega_x \times \mathbb{R}_v^N$, where $\Omega_x = \mathbb{T}^N$ (torus) or Ω_x is a smooth open subset of $\mathbb{R}^N$, strictly convex [nobody really knows what should happen if Ω_x is not convex], then there exists a unique smooth solution $(f_t)_{t \geq 0}$ to the BE with specular reflection or periodic or bounce-back boundary condition, and initial datum f_0, satisfying

$$\sup_{t \geq t_0} \left\| f_t \right\|_{L_s^1} < +\infty \quad \forall s$$

$$\sup_{t \geq t_0} \left\| f_t \right\|_{H^k} < +\infty \quad \forall k$$

$$t \geq t_0 \implies f_t(x, v) \geq \rho_0 e^{-A_0 |v|_0^q}.$$

Here

$$\left\| f \right\|_{L_s^1} = \int f(x, v)(1 + |v|^s) \, dx \, dv,$$

$$\left\| f \right\|_{H^k} = \left\| f \right\|_{H^k(\Omega_x \times \mathbb{R}_v^N)}.$$

At present such a theorem can be deduced only in a close-to-equilibrium context from the recent works by Guo.

The program that I started with Desvillettes some time ago is the following: *assume* that (f_t) satisfies all those conjectured bounds, and get *explicit* rates of convergence for f_t towards global equilibrium. This should be seen as the last step of a complete argument. Of course a compactness proof solves the problem of convergence in a few lines, but gives no information about explicit rates.

The first, naive approach is the following. Start from the H Theorem,

$$\frac{d}{dt} H(f \mid M) = -\int_{\Omega_x} D(ft) dx, \quad (M = \text{ global eq.})$$

and from this deduce that

$$\int_0^{+\infty} D(f_t)\,dx\,dt < +\infty;$$

as a consequence

$$\int_0^{+\infty} H\left(f_t \mid M^{f_t}\right)^{1+\epsilon} dt < +\infty$$

and in particular

$$\int_0^{+\infty} \left\|f_t - M^{f_t}\right\|_{L^1}^{2(1+\epsilon)} dt < +\infty.$$

Thus it seems that f_t approaches local equilibrium, at a rate no more than $O\left(\frac{1}{t^{1/2-0}}\right)$ or something like that.

But this is downright false: A convergent integral may indeed diverge arbitrarily slowly, no matter how smooth the function is: just think of the case where it admits a lot of "plateaux"... (see figure)

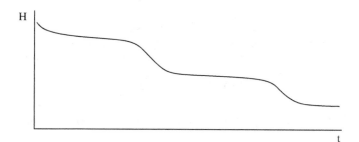

Such a behavior may seem unlikely, but there is some reason to suspect that it can happen; in particular, it is possible to cook up simple linear models, completely integrable, for which the entropy production is oscillating at regular pace. On my request, F. Filbet has performed some very precise numerical simulations on some instances of the spatially inhomogeneous Boltzmann equation in a one-dimensional periodic box (velocity being two-dimensional), and observed beautiful oscillations in the entropy production, etc. As a conclusion, not only is it difficult to rule out the behavior sketched in the above figure, but I personally suspect that it is often the rule! A more precise discussion can be found in Desvillettes and Villani (2005).

From the physical point of view, it is also a heresy to proceed along the lines sketched above. Indeed, contrarily to the hydrodynamic limit, here there is no separation of scales: we are looking at a regime where the mean free path is of the same order as the typical length scale (Knudsen number = $O(1)$). So the right thing to do is to work *at the same time* on the deviation of f from M^f, and on the deviation from M^f to M (or equivalently of ρ, u, T from their equilibrium values). In short:

- Like in the hydrodynamic limit problem we have to study the approximation of M^f by f, and the behavior of ρ, u, T (hydrodynamic fields).
- But on the contrary to the hydrodynamical limit problem, here we should study at the same time the convergence of f towards M^f, and of ρ, u, T towards the equilibrium value. Let's say that the trend to equilibrium problem is at the same time local and global, while the hydrodynamical limit problem is essentially local. I won't discuss this in more detail since both problems are still far to be fully understood.

Theorem 10 (Desvillettes and Villani, 2005). *Let $B(v - v_*, \sigma)$ satisfy*

$$K_B |v - v_*|^\gamma \wedge (1 + |v - v_*|)^{-\beta} \leq B(v - v_*, \sigma) \leq C_B (1 + |v + v_*|)^{k_B} \frac{\theta^{-(1+1/\nu_B)}}{\sin^{N-2} \theta}$$

with $\beta, \gamma \geq 0$, $K_B < +\infty$, $\nu_B < 2$.

Let Ω_x be $\mathbb{T}^N$ or a smooth open subset of $\mathbb{R}^N$, connected, and let $(f_t)_{t \geq 0}$ be a very smooth solution of the BE

$$\frac{\partial f}{\partial f} + v \cdot \nabla_x f = Q(f, f),$$

with periodic boundary conditions or bounce-back reflection or specular reflection. In the case of specular reflection, further assume that Ω has no axis of symmetry and $N = 2$ or 3. Without loss of generality, assume that

$$|\Omega_x| = 1 \quad \int f_0 \, dv \, dx = 1 \quad \int f_0(x, v) |v|^2 \, dv \, dx = N,$$

and if $\Omega_x = \mathbb{T}^N$ assume $\int f_0(x, v) v \, dv \, dx = 0$. Further assume that

$$\forall s \sup_{t \geq 0} \left\| f_t \right\|_{L^1_s} < +\infty$$

$$\forall k \sup_{t \geq 0} \left\| f_t \right\|_{H^k} < +\infty$$

$\exists \rho_0, A_0, q_0; \forall t \geq 0 \quad \forall v \, \forall x \quad f_t(x, v) \geq \rho_0 e^{-A_0 |v|^{q_0}}$.
Then, if M stands for the global equilibrium

$$M(x, v) = M(v) = \frac{e^{-|v|^2/2}}{(2\pi)^{N/2}},$$

one has $H(f_t \mid M) = O(t^{-\infty})$ in the sense that for all $\epsilon > 0$ there exists a constant C_ϵ, only depending on B, Ω_x, on the boundary conditions, on ρ_0, A_0, q_0, on $\sup_{t \geq 0} \|f_t\|_{L^1_s}$ and $\sup_{t \geq 0} \|f_t\|_{H^k}$ for some $s = s(\epsilon, q_0, B)$ and $k = k(\epsilon, q_0, B)$, such that $H(f_t \mid M) \leq C_\epsilon t^{-1/\epsilon}$. In particular

$$\left\| f_t - M \right\|_{L^1(\Omega_x \times \mathbb{R}^N)} = O(t^{-\infty}).$$

I will not reproduce the proof, which includes dozens of pages of "pure computation", but only try to explain the main (hopefully interesting) ideas and the overall plan.

First of all, the general strategy is, as in the spatially homogeneous case, to *reduce to a system of differential inequalities* for which the rate of convergence can be estimated by elementary means.

As long as $H(f \mid M^f)$ [here H is with respect to $\mathrm{d}x\,\mathrm{d}v$] is significantly large with respect to $H(f \mid M)$, one can use the quantitative H Theorem and get

$$-\frac{\mathrm{d}}{\mathrm{d}t}H(f \mid M) = \int_{\Omega_x} D(f_t(x, \cdot))\mathrm{d}x \geq K_{f_t}H(f_t \mid M)^{1+\epsilon}.$$

K constant depending on Ω_x and on bounds on f_t.

This differential inequality, were it true for all times, would ensure $O(t^{-\infty})$ convergence by Growvall's lemma. BUT it may very well happen that $H(f \mid M^f)$ becomes very small at some time, and actually the total entropy production $\int_{\Omega_x} D(f_t(x, \cdot))\mathrm{d}x$ would become very small compared to $H(f \mid M)$.

A general plan of attack against this degeneracy was proposed by Desvillettes and Villani [40]. In the present context it would consist in estimating from below the *second time-derivative* of a well-chosen functional measuring the discrepancy between between f_t and M^{f_t}. In our context the most natural choice would be to look at $\mathrm{d}^2/\mathrm{d}t^2 H(f_t \mid M^{f_t})$; but this leads to very tricky problems to control errors everywhere, and it is much better to look at $\dfrac{\mathrm{d}^2}{\mathrm{d}t^2}\|f_t - M^{f_t}\|_{L^2}^2$. In fact, this is one of the necessary ingredients to avoid using too strong a priori estimates like $f_t \in L^\infty(M^{-1})$ or something like that.

Behind this calculation is the following idea: if, at some time t_*, $f_{t_*} = M^{f_{t_*}}$, then f_t has to strictly depart from M^{f_t} just after t_*, unless $M^{f_{t_*}} = M$ [because M is the only local Maxwellian solving the BE]. Thus the second-order derivative $\mathrm{d}^2/\mathrm{d}t^2\|f_t - M^{f_{t_*}}\|_{L^2}^2$ should be strictly positive whenever f_t is very close to M^{f_t} as ρ_t, u_t, T_t are not at their equilibrium values $(1, 0, 1)$. Then one would combine this differential inequality on $\left\|f_t - M^{f_t}\right\|_{L^2}^2$ with the entropy production estimate (H Theorem) to get a system of differential inequalities involving both the distance between F_t to M^{f_t}, and the distance between M^{f_t} and M.

This strategy was successfully applied to the model case of the linear Fokker–Planck equation (spatially dependent). However, the Boltzmann equation is nastier, and the above guess about $\mathrm{d}^2/\mathrm{d}t^2\|f_t - M^{f_t}\|_{L^2}^2$ is *false*. There are a lot of hydrodynamic fields ρ, u, T (x-dependent!), which are not in equilibrium, and such that this second derivative vanishes. Desvillettes and I called these states *quasi-equilibria*. They correspond to a considerable slowing down of the entropy production: whenever t passes through a quasi-equilibrium at some time t_*, then not only does the total entropy production $\int_{\Omega_x} D(f_{t_*}(x,\cdot))\,\mathrm{d}x$ vanish, but also its time derivative at time t_*; so formally, $H(f_t|M^{f_t}) = O((t-t_*)^4)$ near t_*.

These quasi-equilibria depend on the global geometry of the problem, i.e. both the shape of Ω_x and the boundary conditions. They are determined as the solutions of the equations

$$\begin{bmatrix} T = \text{constant} \\ \{\nabla u\} = 0, \end{bmatrix}$$

where $\{\nabla u\}$ is the deviator of u, i.e. the traceless part of the symmetric gradient of u

$$\{\nabla u\}_{ij} = \frac{1}{2}\left[\frac{\partial u_i}{\partial x_j} + \frac{\partial u_j}{\partial x_i}\right] - \frac{(\nabla.u)}{N}\delta_{ij}.$$

Obviously, for any Ω_x one can construct quasi-equilibria by imposing $T = \text{constant}$, $u = 0$, ρ arbitrary (x-dependent). The existence of solutions with non-trivial velocity field, on the other hand, depends on Ω_x:

- If $\Omega_x = \mathbb{T}^N$, then $\{\nabla u\} = 0 \Longrightarrow \nabla u = 0$ (so $u = 0$, with our normalization).
- If $\Omega_x \subset \mathbb{R}^N$ with bounce-back condition, then $\{\nabla u\} = 0 \Longrightarrow u = 0$.
- If $\Omega_x \subset \mathbb{R}^2$ with specular reflection, then there always exists a 3-dimensional vector space of solution to $\{\nabla u\} = 0$.
- If $\Omega_x \subset \mathbb{R}^3$ with specular reflection, then there is no quasi-equilibrium with non-trivial velocity field unless Ω_x has an axis of symmetry (if Ω_x is a spherical ball, then there is, in addition to the obvious ones, a strange quasi-equilibrium, as shown to me by Ghys).

The remedy which we suggest is to estimate, not only the distance between f and $M(\rho, u, T) = M^f$, but also between f and $M(\rho, u, 1)$, and between f and $M(\rho, 0, 1)$; which is an attempt to separate as much as possible the roles of the three hydrodynamic fields.

Now we have all the main ideas for establishing a system of differential inequalities.

First step: Write a quantitative H Theorem

$$-\frac{\mathrm{d}}{\mathrm{d}t}H(f_t \mid M) = \int_{\Omega_x} D(f_t(x,.))\mathrm{d}x \geq K_1 H(f_t \mid M^{f_t})^{1+\epsilon} \qquad (\text{E.1})$$

Here K_1 depends on f, on B and on ϵ, which will be fixed until the end but arbitrarily small.

Second step: Explicitly compute $\dfrac{\mathrm{d}^2}{\mathrm{d}t^2}\left\|f - M(\rho, u, T)\right\|_{L^2}^2$, $\dfrac{\mathrm{d}^2}{\mathrm{d}t^2}\left\|f - M(\rho, u, 1)\right\|$ $\dfrac{\mathrm{d}^2}{\mathrm{d}t^2}\left\|f - M(\rho, 0, 1)\right\|_{L^2}^2$, where

$$\rho = \rho_t = \int f_t dv, \quad \rho u = \rho_t u_t = \int f_t v dv,$$

$$\rho|u|^2 = \rho_t |u_t|^2 = \frac{1}{N} \int f_t |v - v_t|^2 dv$$

This is a monster computation; for the reader's amusement, I reproduced only the term $\dfrac{1}{M(\rho, u, T)} \dfrac{\partial^2}{\partial t^2} M(\rho, u, T)$ at the very end of the notes. This expression may look frightening, but from the formula given one sees that it is formally $O(\|f - M\|)$ and this is all one has to retain from this intermediate computation. After one has separated the important contributions from the rest, one arrives at

$$\frac{d^2}{dt^2} \left\| f - M(\rho, u, T) \right\|_{L^2}^2 \geq K_2 \left[\int_{\Omega_x} \{\nabla u\}^2 + \int_{\Omega_x} |\nabla T|^2 \right]$$

$$- \frac{C_2}{\delta_2^{1-\epsilon}} \left\| f - M(\rho, u, T) \right\|_{L^2}^{2(1-\epsilon)} - \delta_2 H(f \mid M) \tag{E.2}$$

Here δ_2 is arbitrarily small, and the constants depend on the a priori estimates on f (and thus on $M(\rho, u, T)$ as well). The appearance of the ϵ in the second term of the right-hand side is due to an *interpolation* procedure, which is mandatory if one wants to measure the distance between f and $M(\rho, u, T)$ with the L^2 norm. Indeed, very strong norms (with derivatives with respect to v and x) of $f - M(\rho, u, T)$ appear in the process of estimating $\dfrac{d^2}{dt^2} \| f - M(\rho, u, T)\|_{L^2}^2$; then to "transform" these strong norms in L^2, one interpolates with the smoothness bounds on f and $M(\rho, u, T)$, and an ϵ exponent is lost in the battle. Note that the lower bound assumption on f is important here to ensure the smoothness of ρ, u and T (and the strict positivity of ρ and T).

Also note the asymmetry between the way we quantify the distance from f to $M(\rho, u, T)$, and the distance from f to M. Using $H(f \mid M(\rho, u, T))$ would require much stronger assumptions on the tail behavior of f.

Equation (E.2) can interpreted as follows: if the distance between f and $M^f = M(\rho, u, T)$, ever becomes extremely close to 0, and if $\{\nabla u\} \neq 0$ or $\nabla T \neq 0$ [so that $M(\rho, u, T)$ is not a quasi-equilibrium] then, up to a small error $(\delta_2 H(f \mid M))$ the rate at which f_t will depart from M^{f_t} at later times is estimated from below.

One establishes in a similar way

$$\frac{d^2}{dt^2} \left\| f - M(\rho, u, 1) \right\|_{L^2}^2 \geq K_3 \int_{\Omega_x} |\nabla^{\mathrm{sym}} u|^2 - \frac{C_3}{\delta_3^{1-\epsilon}} \left\| f - M(\rho, u, 1) \right\|_{L^2}^{2(1-\epsilon)}$$

$$- \delta_3 H(f \mid M), \tag{E.3}$$

$$\frac{d^2}{dt^2}\left\|f - M(\rho, 0, 1)\right\|_{L^2}^2 \geq K_4\left(\int_{\Omega_x} |\nabla\rho|^2 + \int_{\Omega_x} (\nabla_x u)^2\right)$$
$$- \frac{C_4}{\delta_4^{1-\epsilon}}\left\|f - M(\rho, u, 1)\right\|_{L^2}^{2(1-\epsilon)} - \delta_4 H(f \mid M). \quad \text{(E.4)}$$

Here $\nabla^{\mathrm{sym}} u$ is the symmetrized gradient of u:

$$\left[\nabla^{\mathrm{sym}} u\right]_{ij} = \frac{\partial_i u_j + \partial_j u_i}{2}$$

Note that on the whole, the quantities $\int |\nabla\rho|^2$, $\int |\nabla^{\mathrm{sym}} u|^2$, $\int |\nabla T|^2$ have appeared, and that the combination of all of them controls how far ρ, u, T are from equilibrium. This actually is the content of

Step 3: One establishes *functional inequalities* to quantify how the quadratic ∇_x quantities seen above control the distance between the hydrodynamic fields and their equilibrium values. At this point one uses

- The lower bound on ρ
- The boundary conditions
- The shape of Ω_x
- The *global* conservation laws

Remark. The fact that these quadratic gradient quantities appear is a consequence of our strategy which looks at second derivatives along the differential operator $v.\nabla_x$, in such a way that the square of the partial derivative $\frac{\partial}{\partial t}$ contributes to the main part.

- *Inequality for ρ.* Under the constraint $\int \rho = 1$, $|\Omega_x| = 1$, one has

$$\int_{\Omega_x} |\nabla\rho|^2 \geq K_5 \int \rho \log \rho \quad \text{(Poincaré inequality in fact)} \quad \text{(E.5)}$$

- *Inequality for u.* If $\Omega_x = \mathbb{T}^N$ or $\Omega_x \subset \mathbb{R}^N$ with bounce-back condition ($u = 0$ on $\partial\Omega_x$) or $\Omega_x \subset \mathbb{R}^N$ with specular reflection ($u \cdot n = 0$ on $\partial\Omega_x$) with no axis of symmetry, then

$$\int_{\Omega_x} |\nabla^{\mathrm{sym}} u|^2 \geq K_6 \int_{\Omega_x} |\nabla u|^2 \quad \text{(E.6)}$$

This is a variant of *Korn's inequality*, well-known in elasticity and in hydrodynamics. One original thing here is that it appears with the tangency constraint $u \cdot n = 0$, while in both above-mentioned fields it usually comes with no slip boundary conditions: $u = 0$ on some portion of $\partial\Omega_x$. The constant K_6 can be explicitly estimated in terms of how much the domain Ω_x departs from being axisymmetric.

Next, by Poincaré's inequality *with reference measure* $\rho(x)dx$, [which is uniformly smooth and bounded from below, and has unit integral!]

$$\int_{\Omega_x} |\nabla u|^2 \geq \frac{1}{\|\rho\|_{L^\infty}} \int \rho |\nabla u|^2 \geq \frac{K_6'}{\|\rho\|_{L^\infty}} \int \rho |u|^2. \qquad (\text{E.6'})$$

(here we use the boundary condition on u again; and $\int \rho u = 0$ if $\Omega_x = \mathbb{T}^N$.)

- *Inequality for T.* Using the fact that ρ is bounded from below and above,

$$\int \rho T + \frac{1}{N} \int \rho |u|^2 = 1,$$

and a Poincaré inequality, one shows

$$\int_{\Omega_x} |\nabla T|^2 \geq K_7 \times \frac{N}{2} \int \rho(T - \log T - 1) \, dx - C_7 \int \rho \frac{|u|^2}{2} \, dx \qquad (\text{E.7})$$

Our strange choices for estimating the distance of ρ, u, T to $1, 0, 1$ will become clear after the next step.

Step 4: It consists in quantifying the idea that if f is close to M^f, and M^f is close to M, then f is close to M. This is provided by a well-known formula of *additivity of the entropies*:

$$H(f \mid M) \qquad\qquad\qquad\qquad\qquad\qquad\qquad\qquad (\text{E.8})$$

$$= H(f \mid M(\rho, u, T)) + \left[\int \rho \log \rho + \int \rho \frac{|u|^2}{2} + \frac{N}{2} \int \rho(T - \log T - 1) \right]$$

Let us complement it with the similar formulas

$$H(f \mid M) = H(f \mid M(\rho, u, 1)) + \left[\int \rho \log \rho + \int \rho \frac{|u|^2}{2} \right] \qquad (\text{E.9})$$

$$H(f \mid M) = H(f \mid M(\rho, 0, 1)) + \int \rho \log \rho \qquad (\text{E.10})$$

Equations (E.1)–(E.10) constitute a system of differential inequalities involving the quantities $H(f \mid M)$, $H(f \mid M(\rho, u, T))$, $H(f \mid M(\rho, u, 1))$, $H(f \mid M(\rho, 0, 1))$, $\|f - M(\rho, u, T)\|_{L^2}^2$, $\|f - M(\rho, u, 1)\|_{L^2}^2$, $\|f - M(\rho, 0, 1)\|_{L^2}^2$. To reduce the number of quantities involved, we can use interpolation again:

$$\left\| f - M(\rho, u, T) \right\|_{L^2}^2 \leq C_{11} H(f \mid M(\rho, u, T))^{1-\epsilon} \qquad (\text{E.11})$$

$$\left\| f - M(\rho, u, 1) \right\|_{L^2}^2 \leq C_{12} H(f \mid M(\rho, u, 1))^{1-\epsilon} \qquad \text{(E.12)}$$

$$\left\| f - M(\rho, 0, 1) \right\|_{L^2}^2 \leq C_{13} H(f \mid M(\rho, 0, 1))^{1-\epsilon}. \qquad \text{(E.13)}$$

Now we have a closed system of differential inequalities on the quantities $H(f \mid M)$, $\|f - M(\rho, u, T)\|_{L^2}^2$, $\|f - M(\rho, u, 1)\|_{L^2}^2$, $\|f - M(\rho, 0, 1)\|_{L^2}^2$. To study its time behavior, we need to quantify the fact that a solution of

$$y''(t) \geq K - Cy(t)^{1-\epsilon}$$

cannot be always small if $K > 0$. For this we make systematic use of the following lemma from Desvillettes and Villani [40]:

Lemma L. 1. *Let $y(t) \geq 0$ be a C^2 function solving the differential inequality*

$$\frac{\mathrm{d}^2}{\mathrm{d}t^2} y(t) + \frac{CA}{\delta} y^{1-\epsilon} \geq K\alpha \quad \text{for } T_1 \leq t \leq T_2$$

where C, A, δ, K, α are positive constants, A bounded below, ϵ small enough. Then,

- *Either $[T_1, T_2]$ is short:*

$$T_2 - T_1 \leq \text{const.} \left(\frac{A}{\delta} \right)^{-\frac{1}{2(1-\epsilon)}} \alpha^{\frac{\epsilon}{2(1-\epsilon)}}$$

- *Or $y(t)$ is bounded below on the average:*

$$\frac{1}{T_2 - T_1} \int_{T_1}^{T_2} y(t) \, \mathrm{d}t \geq \text{const} \left(A^{-1} \delta \right)^{\frac{3}{2} + 10\epsilon} \alpha^{1/1-\epsilon}.$$

with constants only depending on C, K and ϵ (not on A, δ, α).

In short: either $T_2 - T_1$ is bounded in $O(\alpha^0)$ or the average value of y is bounded below in $O(\alpha^{1+0})$.

The dichotomy has to do with the fact that we are dealing with *second-order* differential inequalities, for which there is no such thing as Gronwall's lemma, and there may be some *delay* before the positivity of the right-hand side is "transferred" onto $y(t)$. One can easily see that anything can happen if $[T_1, T_2]$ is too short.

The lemma is established "by hand" and relies on a careful cutting of $[T_1, T_2]$ into many sub-pieces depending on the values take by $y(t)$. the main idea is that if y becomes very small, then, due to the inequality, it has to become strictly convex as a function of t, then either it will very soon leave this regime where it is very small, or it will stay very small for a long time, but then, due to uniform convexity, finally get out of the smallness regime with a very fast rate of increase.

If we want to apply Lemma L.1 to the differential inequalities established in previous steps, we run into a serious difficulty: we would like to use (E.2) only when $\int |\nabla T|^2$ (or $\int \rho(T - \log T - 1)$) is large and $H(f \mid M(\rho, u, T))$ is very small; (E.3) only when $\int \rho |u|^2$ is large and $H(f \mid M(\rho, u, 1))$ is very small; (E.4) only when $\int \rho \log \rho$ is large and $H(f \mid M(\rho, 0, 1))$ is very small. In each case, we need lower bounds on the lengths of the time intervals on which each event occurs. What if f_t switches very quickly from one of these "bad" states to the other, and keeps doing that? Then we cannot apply Lemma L.1 in an interesting way. So the next step consists in ruling out this possibility, by establishing that the hydrodynamic quantities vary "slowly" in term:

Step 5: By using the Boltzmann equation again, together with interpolation and lower bounds, one establishes that

$$\left| \frac{\mathrm{d}}{\mathrm{d}t} \int \rho_t \log \rho_t, \; \frac{\mathrm{d}}{\mathrm{d}t} \int \rho_t |u_t|^2, \; \frac{\mathrm{d}}{\mathrm{d}t} \int \rho_t (T_t - \log T_t - 1) \right| \leq C_{14} H(f \mid M)^{1-\epsilon} \tag{E.14}$$

(when the system approaches global equilibrium the hydrodynamic quantities vary slowly.)

Note that we do *not* know how to establish a similar bound for $\left| \frac{\mathrm{d}}{\mathrm{d}t} H(f_t) \right|$ [kinetic entropy!] without stronger assumptions on the tail behavior of f_t (pointwise exponential bounds or so).

All the functional job has now been done and we are left with a problem on differential inequalities (which does not mean that the rest is easy......). The last step will yield the conclusion:

Step 6: Using the fact that the quantities

$$H(f_t \mid M), \quad \|f_t - M(\rho, u, T)\|_{L^2}^2, \quad \|f_t - M(\rho, u, 1)\|_{L^2}^2, \quad \|f_t - M(\rho, 0, 1)\|_{L^2}$$

and the quantities $\int \rho_t \log \rho_t, \; \int \rho_t |u_t|^2, \; \frac{N}{2} \int \rho_t (T_t - \log T_t - 1)$ satisfy the system of differential inequalities given by (E.1)–(E.14), and are bounded quantities, prove that

$$H(f_t \mid M) = O(t^{-1/250\epsilon}).$$

This step is again done by cutting into sub-pieces a time-interval on which $\alpha \geq H(f_t \mid M) \geq \lambda \alpha$, where $\lambda < 1$ is fixed ($\lambda = 4/5$), such that on the sub-pieces the entropy production is either high or small. Then re-cut into pieces the intervals on which the entropy production is small, according to whether $\int |\nabla T|^2$ is large or small; then repeat the process for $\int \rho |u|^2$, then again for $\int \rho \log \rho \dots$. Then use all the differential inequalities, together with lemma L.1, to show a lower bound *on the average* on the entropy production, and in the end to bound the length of the time interval into consideration by a function of α.

1.8 Towards Exponential Convergence?

For various reasons, our method of proof cannot yield exponential decay. In certain situations it is not expected that exponential decay occurs: in particular, for soft potentials, the linearized Boltzmann operator has no spectral gap and no exponential convergence is expected (Caflisch [14] has some results in $O(e^{-t^\alpha})$ for some cases). However, there are other cases, such as hard spheres, in which exponential decay is expected. There is no hope in trying to fix our proof to get such a result, it is much better to turn to a *linearized* study. One of our goals when working on this problem was to find a method which does not use linearization, mainly because linearization is not so general and does not yield any control on the time it takes until one enters the linearized regime. But now that we have some quantitative estimates, we can hope to complement them with a linearized study.

A major difficulty arising here is that our setting is not the natural one for the linearization: to get a self-adjoint linear Boltzmann collision operator, the natural space is $L^2(M^{-1})$, which is much, much narrower than the spaces we have been working in. It looks extremely doubtful that we can establish strong enough estimates to reduce to that space, so the best thing to do is accept the idea that we are going to work in a non-self-adjoint setting. Such a study was already performed by Arkeryd [4], with later refinements due to Wennberg [90,91]. Still, their work was limited to the spatially homogeneous case, and also involved a compactness argument to reduce from the self-adjoint to the non-self-adjoint case. Moreover, spectral gap in the self-adjoint case itself was based on a standard compactness argument (a consequence of Weyl's perturbation theorem for linear operators), thus preventing any hope of getting explicit estimates whatsoever.

So here is the program:

1. Find a constructive method for bounding below the spectral gap in the self-adjoint case, i.e. in the natural space $L^2(M^{-1})$, say for hard spheres;
2. Find a constructive argument to go from spectral gap in $L^2(M^{-1})$ to spectral gap in L^1, with all the subtleties associated with spectral theory of non-self-adjoint operators in infinite dimension...

3. Find a constructive argument to overcome the degeneracy in the space variable, to get an exponential decay for the linear semigroup associated with the linearized spatially inhomogeneous Boltzmann equation; something similar to hypo-ellipticity techniques.
4. Combine the whole thing with a perturbative and linearization analysis to get the exponential decay for the nonlinear equation, very close to equilibrium.

This of course is a long-run program, on which my student and collaborator Mouhot has been able to make substantial progress, in collaboration with Baranger on one hand, Gallay on the other hand. As of now, Step 1 has been solved in [6] with a clever method based on intermediate collisions. The reader can consult their paper for a very clear exposition. In a much simpler context, this method of intermediate collisions was used by Carrillo et al. [30]; and in a different study, I had tried once to use intermediate collisions to improve some tricky regularity estimates for the Boltzmann equation with grazing collisions [81]; moreover, I recently learnt from Diaconis that the main idea is very close to a useful trick in the study of Markov kernels. In spite of all these precursors, the proof by Baranger and Mouhot is extremely original and, in my opinion, the physical idea which underlies it is quite appealing.

In relation with this, Mouhot and Strain [63] recently clarified the role of the grazing collisions (non-cutoff assumption) in the spectral theory of the linearized Boltzmann equation; a topic which had been studied by Pao and Klaus several decades ago.

Now for the rest of the program: Steps 2 and 4 have been worked out by Mouhot [62] in the case of spatially homogeneous Boltzmann equation with hard potentials; one of the basic ingredients comes from an earlier work by Gallay and Wayne [45], but required significant effort to adapt to the Boltzmann case. Up to refinements and improvements, it seems fair to state that the spatially homogeneous situation is now understood.

All the final effort should now bear on Step 3. I have worked on that recently, developing estimates for various linear operators, in a way which is somewhat reminiscent of hypo-ellipticity but has a distinct flavor – a phenomenon which I call "hypo-coercivity", following a suggestion by Gallay. Many other people have been recently working on that problem with various tools and for various inhomogeneous linear Fokker–Planck equations: for instance Talay [72],Rey-Bellet and Thomas [68], Mattingly and co-workers [58, 59], using ergodic Markov chain theory; Hérau and Nier [49] and Eckmann and Hairer [42], using simple pseudo-differential theory. The work by Hérau and Nier was actually the first one in which an explicit exponential rate was derived for an inhomogeneous Fokker–Planck equation; a recent volume of Lecture Notes by Hérau and Nier [48] gives a simplified and more general variant of their approach, together with an overview of the field. The typical example of such an equation is the following:

$$\frac{\partial f}{\partial t} + v \cdot \nabla_x f - \nabla V(x) \cdot \nabla_v f = \nabla_v \cdot (\nabla_v f + fv).$$

As the reader can see, the left-hand side is transport and confinement by a potential, while the right-hand side is a standard linear Fokker–Planck operator, which contains drift and diffusion, BUT only in the velocity variable.

In this problem (at least in kinetic theory) an analytic approach has in the end many advantages over a stochastic approach, one of the reasons being that most kinetic equations have a very complicated stochastic interpretation; they cannot be written just as the law of diffusion process as is the case for the inhomogeneous Fokker–Planck equation. Note that the nonlinear method described in the previous chapter has been applied to the very same linear inhomogeneous Fokker–Planck equation by Desvillettes and Villani [40], only to get an $O(t^{-\infty})$ rate of convergence.

All my recent progress on this question is gathered in a preprint (*Hypocoercivity*, 2006) which will appear as a Memoir of the American Mathematical Society.

The story is not over and clearly it will soon get into new developments. For instance I would bet that the exponential convergence for the inhomogeneous Boltzmann equation, conditional to strong regularity estimates, can be obtained in the next few years.

References

1. Alexandre, R. and Villani, C. (2004). On the Landau approximation in plasma physics. *Ann. Inst. H. Poincaré Anal. Non Linéaire*, 21(1):61–95.
2. Ané, C., Blachère, S., Chafaï, D., Fougères, P., Gentil, I., Malrieu, F., Roberto, C., and Scheffer, G. (2000). *Sur les inégalités de Sobolev logarithmiques*, volume 10 of *Panoramas et Synthèses [Panoramas and Syntheses]*. Société Mathématique de France, Paris. With a preface by Dominique Bakry and Michel Ledoux.
3. Arkeryd, L. (1983). L^∞ estimates for the space-homogeneous Boltzmann equation. *J. Statist. Phys.*, 31(2):347–361.
4. Arkeryd, L. (1988). Stability in L^1 for the spatially homogeneous Boltzmann equation. *Arch. Rational Mech. Anal.*, 103(2):151–167.
5. Ball, K., Barthe, F., and Naor, A. (2003). Entropy jumps in the presence of a spectral gap. *Duke Math. J.*, 119(1):41–63.
6. Baranger, C. and Mouhot, C. (2005). Explicit spectral gap estimates for the linearized Boltzmann and Landau operators with hard potentials. *Rev. Mat. Iberoamericana*, 21(3):819–841.
7. Blachman, N. M. (1965). The convolution inequality for entropy powers. *IEEE Trans. Inform. Theory*, 2:267–271.
8. Bobylev, A. V. (1988). The theory of the nonlinear, spatially uniform Boltzmann equation for Maxwellian molecules. *Sov. Sci. Rev. C. Math. Phys.*, 7:111–233.
9. Bobylev, A. V. and Cercignani, C. (1999). On the rate of entropy production for the Boltzmann equation. *J. Statist. Phys.*, 94(3-4):603–618.
10. Bobylev, A. V. and Cercignani, C. (2002a). Exact eternal solutions of the Boltzmann equation. *J. Statist. Phys.*, 106(5-6):1019–1038.

11. Bobylev, A. V. and Cercignani, C. (2002b). The inverse Laplace transform of some analytic functions with an application to the eternal solutions of the Boltzmann equation. *Appl. Math. Lett.*, 15(7):807–813.
12. Bobylev, A. V. and Cercignani, C. (2002c). Self-similar solutions of the Boltzmann equation for non-Maxwell molecules. *J. Statist. Phys.*, 108(3-4):713–717.
13. Bonami, A. (1970). Étude des coefficients de Fourier des fonctions de $L^p(G)$. *Ann. Inst. Fourier (Grenoble)*, 20(fasc. 2):335–402 (1971).
14. Caflisch, R. (1980). The Boltzmann equation with a soft potential. *Comm. Math. Phys.*, 74:71–109.
15. Caputo, P. (2003). Uniform Poincaré inequalities for unbounded conservative systems: the non-interacting case. *Stochastic Process. Appl.* **106**, no.2, pp. 223–244.
16. Carleman, T. (1932). Sur la théorie de l'equation intégrodifférentielle de Boltzmann. *Acta Math.*, 60:369–424.
17. Carlen, E. A. (1991a). Some integral identities and inequalities for entire functions and their application to the coherent state transform. *J. Funct. Anal.*, 97(1):231–249.
18. Carlen, E. A. (1991b). Superadditivity of Fisher's information and logarithmic Sobolev inequalities. *J. Funct. Anal.*, 101(1):194–211.
19. Carlen, E. A. and Carvalho, M. C. (1992). Strict entropy production bounds and stability of the rate of convergence to equilibrium for the Boltzmann equation. *J. Statist. Phys.*, 67(3-4):575–608.
20. Carlen, E. A. and Carvalho, M. C. (1994). Entropy production estimates for Boltzmann equations with physically realistic collision kernels. 74(3-4):743–782.
21. Carlen, E. A., Carvalho, M. C., and Gabetta, E. (2000). Central limit theorem for Maxwellian molecules and truncation of the Wild expansion. *Comm. Pure Appl. Math.*, 53(3):370–397.
22. Carlen, E. A., Carvalho, M. C., and Loss, M. (2001). Many-body aspects of approach to equilibrium. In *Séminaire: Équations aux Dérivées Partielles, 2000–2001*, Sémin. Équ. Dériv. Partielles, pages Exp. No. XIX, 12. École Polytech., Palaiseau.
23. Carlen, E. A., Carvalho, M. C., and Loss, M. (2003). Determination of the spectral gap for Kac's master equation and related stochastic evolution. *Acta Math.*, 191(1):1–54.
24. Carlen, E. A., Carvalho, M. C., and Wennberg, B. (1997). Entropic convergence for solutions of the Boltzmann equation with general physical initial data. *Transport Theory Statist. Phys.*, 26(3):373–378.
25. Carlen, E. A., Esposito, R., Lebowitz, J. L., Marra, R., and Rokhlenko, A. (1998). Kinetics of a model weakly ionized plasma in the presence of multiple equilibria. *Arch. Rational Mech. Anal.*, 142(3):193–218.
26. Carlen, E. A., Ester, G., and Regazzini, E. (2007). On the rate of explosion for infinite energy solutions of the spatially homogeneous Boltzmann equation. Preprint.
27. Carlen, E. A., Gabetta, E., and Toscani, G. (1999). Propagation of smoothness and the rate of exponential convergence to equilibrium for a spatially homogeneous Maxwellian gas. *Comm. Math. Phys.*, 199(3):521–546.
28. Carlen, E. A. and Lu, X. (2003). Fast and slow convergence to equilibrium for Maxwellian molecules via Wild sums. *J. Statist. Phys.*, 112(1-2):59–134.
29. Carlen, E. A. and Soffer, A. (1991). Entropy production by block variable summation and central limit theorems. *Comm. Math. Phys.*, 140:339–371.
30. Carrillo, J. A., McCann, R. J., and Villani, C. (2003). Kinetic equilibration rates for granular media and related equations: entropy dissipation and mass transportation estimates. *Rev. Mat. Iberoamericana*, 19(3):971–1018.
31. Cercignani, C. (1982). *H*-theorem and trend to equilibrium in the kinetic theory of gases. *Arch. Mech.*, 34:231–241.
32. Cover, T. M. and Thomas, J. A. (1991). *Elements of information theory*. Wiley Series in Telecommunications. John Wiley & Sons Inc., New York. A Wiley-Interscience Publication.
33. Dembo, A. (1989). Simple proof of the concavity of the entropy power with respect to added Gaussian noise. *IEEE Trans. Inform. Theory*, 35(4):887–888.

34. Dembo, A., Cover, T., and Thomas, J. (1991). Information theoretic inequalities. *IEEE Trans. Inform. Theory*, 37(6):1501–1518.

35. Desvillettes, L. (1989). Entropy dissipation rate and convergence in kinetic equations. *Comm. Math. Phys.*, 123(4):687–702.

36. Desvillettes, L. (1990). Convergence to equilibrium in large time for Boltzmann and BGK equations. *Arch. Rational Mech. Anal.*, 110(1):73–91.

37. Desvillettes, L. (1993). Some applications of the method of moments for the homogeneous Boltzmann equation. *Arch. Rational Mech. Anal.*, 123(4):387–395.

38. Desvillettes, L. (2000). Convergence to the thermodynamical equilibrium. In G. Iooss, O. G. and Nouri, A., editors, *Trends in Applications of Mathematics to Mechanics, Monographs and Surveys in Pure and Applied Mathematics*, pages 115–126. Chapman and Hall, Boca Raton.

39. Desvillettes, L. and Villani, C. (2000). On the spatially homogeneous Landau equation for hard potentials. II. *H*-theorem and applications. *Comm. Partial Differential Equations*, 25(1-2):261–298.

40. Desvillettes, L. and Villani, C. (2001). On the trend to global equilibrium in spatially inhomogeneous entropy-dissipating systems: the linear Fokker-Planck equation. *Comm. Pure Appl. Math.*, 54(1):1–42.

41. Desvillettes, L. and Villani, C. (2005). On the trend to global equilibrium for spatially inhomogeneous kinetic systems: the Boltzman equation. *Invent. Math.*, 159(2): 245–316.

42. Eckmann, J.-P. and Hairer, M. (2003). Spectral properties of hypoelliptic operators. *Comm. Math. Phys.*, 235(2):233–253.

43. Fisher, R. A. (1925). Theory of statistical estimation. *Math. Proc. Cambridge Philos. Soc.*, 22:700–725.

44. Gabetta, E., Toscani, G., and Wennberg, B. (1995). Metrics for probability distributions and the trend to equilibrium for solutions of the Boltzmann equation. *J. Statist. Phys.*, 81:901–934.

45. Gallay, T. and Wayne, C. E. (2002). Invariant manifolds and the long-time asymptotics of the Navier-Stokes and vorticity equations on $\mathbf{R}^2$. *Arch. Ration. Mech. Anal.*, 163(3):209–258.

46. Gustafsson, T. (1986). L^p-estimates for the nonlinear spatially homogeneous Boltzmann equation. *Arch. Rational Mech. Anal.*, 92(1):23–57.

47. Gustafsson, T. (1988). Global L^p-properties for the spatially homogeneous Boltzmann equation. *Arch. Rational Mech. Anal.*, 103:1–38.

48. Helffer, B. and Nier, F. (2005). *Hypoelliptic estimates and spectral theory for Fokker-Planck operators and Witten Laplacians*, volume 1862 of *Lecture Notes in Mathematics*. Springer-Verlag, Berlin.

49. Hérau, F. and Nier, F. (2004). Isotropic hypoellipticity and trend to equilibrium for the Fokker-Planck equation with a high-degree potential. *Arch. Ration. Mech. Anal.*, 171(2):151–218.

50. Janvresse, E. (2001). Spectral gap for Kac's model of Boltzmann equation. *Ann. Probab.*, 29(1):288–304.

51. Kac, M. (1956). Foundations of kinetic theory. In *Proceedings of the Third Berkeley Symposium on Mathematical Statistics and Probability, 1954–1955, vol. III*, pages 171–197, Berkeley and Los Angeles. University of California Press.

52. Ledoux, M. (1995). L'algèbre de Lie des gradients itérés d'un générateur markovien – développements de moyennes et d'entropies. *Ann. Scient. Ec. norm. sup.*, 28:435–460.

53. Lieb, E. (1982). Comment on : "Approach to Equilibrium of a Boltzmann-Equation Solution". *Phys. Rev. Lett.*, 48(15):1057.

54. Lieb, E. H. (1978). Proof of an entropy conjecture of Wehrl. *Comm. Math. Phys.*, 62(1):35–41.

55. Linnik, Y. V. (1959). An information-theoretic proof of the central limit theorem with the Lindenberg condition. *Theory Probab. Appl.*, 4:288–299.

56. Lions, P.-L. (1994). Compactness in Boltzmann's equation via Fourier integral operators and applications, I. *J. Math. Kyoto Univ.*, 34(2):391–427.
57. Maslen, D. K. (2003). The eigenvalues of Kac's master equation. *Math. Z.*, 243(2):291–331.
58. Mattingly, J. C. and Stuart, A. M. (2002). Geometric ergodicity of some hypo-elliptic diffusions for particle motions. *Markov Process. Related Fields*, 8(2):199–214. Inhomogeneous random systems (Cergy-Pontoise, 2001).
59. Mattingly, J. C., Stuart, A. M., and Higham, D. J. (2002). Ergodicity for SDEs and approximations: locally Lipschitz vector fields and degenerate noise. *Stochastic Process. Appl.*, 101(2):185–232.
60. McKean, H. J. (1966). Speed of approach to equilibrium for Kac's caricature of a Maxwellian gas. *Arch. Rational Mech. Anal.*, 21:343–367.
61. Mischler, S. and Wennberg, B. (1999). On the spatially homogeneous Boltzmann equation. *Ann. Inst. H. Poincaré Anal. Non Linéaire*, 16(4):467–501.
62. Mouhot, C. (2006). Rate of convergence to equilibrium for the spatially homogeneous Boltzmann equation with hard potentials. *Comm. Math. Phys.*, 261(3):629–672.
63. Mouhot, C. and Strain, R. (2006). Spectral gap and coercivity estimates for the linearized boltzmann collision operator without angular cutoff. Preprint.
64. Mouhot, C. and Villani, C. (2004). Regularity theory for the spatially homogeneous Boltzmann equation with cut-off. *Arch. Ration. Mech. Anal.*, 173(2):169–212.
65. Olaussen, K. (1982). Extension of the Boltzmann H theorem. *Phys. Rev. A*, 25(6):3393–3395.
66. Plastino, A. R. and Plastino, A. (1996). Symmetries of the fokker-planck equation and the fisher-frieden arrow time. *Phys. Rev. E*, 54:4423.
67. Pulvirenti, A. and Wennberg, B. (1997). A Maxwellian lower bound for solutions to the Boltzmann equation. *Comm. Math. Phys.*, 183:145–160.
68. Rey-Bellet, L. and Thomas, L. E. (2000). Asymptotic behavior of thermal nonequilibrium steady states for a driven chain of anharmonic oscillators. *Comm. Math. Phys.*, 215(1):1–24.
69. Shannon, C. E. (1948). A mathematical theory of communication. *Bell System Tech. J.*, 27:379–423, 623–656.
70. Shannon, C. E. and Weaver, W. (1949). *The Mathematical Theory of Communication*. The University of Illinois Press, Urbana, Ill.
71. Stam, A. (1959). Some inequalities satisfied by the quantities of information of Fisher and Shannon. *Inform. Control*, 2:101–112.
72. Talay, D. (2002). Stochastic Hamiltonian systems: exponential convergence to the invariant measure, and discretization by the implicit Euler scheme. *Markov Process. Related Fields*, 8(2):163–198. Inhomogeneous random systems (Cergy-Pontoise, 2001).
73. Tanaka, H. (1973). An inequality for a functional of probability distributions and its application to Kac's one-dimensional model of a Maxwellian gas. *Z. Wahrscheinlichkeitstheorie und Verw. Gebiete*, 27:47–52.
74. Tanaka, H. (1978). Probabilistic treatment of the Boltzmann equation of Maxwellian molecules. *Z. Wahrsch. Verw. Gebiete*, 46(1):67–105.
75. Toscani, G. (1992a). Lyapunov functionals for a Maxwell gas. *Arch. Rational Mech. Anal.*, 119:301–307. *NOTA: The argument used to apply the main result in this paper to the Fisher information (a.k.a. Linnik functional) is wrong in dimension 3. See Villani (1998) for a (hopefully) correct proof.*
76. Toscani, G. (1992b). New a priori estimates for the spatially homogeneous Boltzmann equation. *Cont. Mech. Thermodyn.*, 4:81–93.
77. Toscani, G. (1999). Entropy production and the rate of convergence to equilibrium for the Fokker-Planck equation. *Quart. Appl. Math.*, 57(3):521–541.
78. Toscani, G. and Villani, C. (1999a). Probability metrics and uniqueness of the solution to the Boltzmann equation for a Maxwell gas. *J. Statist. Phys.*, 94(3–4):619–637.

79. Toscani, G. and Villani, C. (1999b). Sharp entropy dissipation bounds and explicit rate of trend to equilibrium for the spatially homogeneous Boltzmann equation. *Comm. Math. Phys.*, 203(3):667–706.

80. Villani, C. (1998). Fisher information bounds for Boltzmann's collision operator. *J. Math. Pures Appl.*, 77:821–837.

81. Villani, C. (1999). Regularity estimates via the entropy dissipation for the spatially homogeneous Boltzmann equation without cut-off. *Rev. Mat. Iberoamericana*, 15(2):335–352.

82. Villani, C. (2000a). Contribution à l'étude mathématique des collisions en théorie cinétique. Master's thesis, Univ. Paris-Dauphine, France.

83. Villani, C. (2000b). Decrease of the Fisher information for solutions of the spatially homogeneous Landau equation with Maxwellian molecules. *Math. Models Methods Appl. Sci.*, 10(2):153–161.

84. Villani, C. (2000c). A short proof of the "concavity of entropy power". *IEEE Trans. Inform. Theory*, 46(4):1695–1696.

85. Villani, C. (2002a). Limites hydrodynamiques de l'équation de Boltzmann (d'après C. Bardos, F. Golse, C. D. Levermore, P.-L. Lions, N. Masmoudi, L. Saint-Raymond). *Astérisque*, (282):Exp. No. 893, ix, 365–405. Séminaire Bourbaki, Vol. 2000/2001.

86. Villani, C. (2002b). A review of mathematical topics in collisional kinetic theory. In *Handbook of mathematical fluid dynamics, Vol. I*, pages 71–305. North-Holland, Amsterdam.

87. Villani, C. (2003). Cercignani's conjecture is sometimes true and always almost true. *Comm. Math. Phys.*, 234(3):455–490.

88. Villani, C. (2006). Hypocoercivity. Preprint, available online via www.umpa.ens-lyon.fr/~cvillani.

89. Wennberg, B. (1992). On an entropy dissipation inequality for the Boltzmann equation. *C.R. Acad. Sci. Paris, Série I,*, 315:1441–1446.

90. Wennberg, B. (1993a). Stability and exponential convergence in L^p for the spatially homogeneous Boltzmann equation. *Nonlinear Anal.*, 20(8):935–964.

91. Wennberg, B. (1993b). Stability and exponential convergence in L^p for the spatially homogeneous Boltzmann equation. *Nonlinear Anal.*, 20(8):935–964.

92. Wennberg, B. (1994a). On moments and uniqueness for solutions to the space homogeneous Boltzmann equation. *Transport Theory Statist. Phys.*, 24(4):533–539.

93. Wennberg, B. (1994b). Regularity in the Boltzmann equation and the Radon transform. *Comm. Partial Differential Equations*, 19(11-12):2057–2074.

94. Wennberg, B. (1995). Stability and exponential convergence for the Boltzmann equation. *Arch. Rational Mech. Anal.*, 130(2):103–144.

95. Wennberg, B. (1996). The Povzner inequality and moments in the Boltzmann equation. In *Proceedings of the VIII International Conference on Waves and Stability in Continuous Media, Part II (Palermo, 1995)*, number 45, part II, pages 673–681.

96. Wennberg, B. (1997). Entropy dissipation and moment production for the Boltzmann equation. *J. Statist. Phys.*, 86(5–6):1053–1066.

APPENDIX

$$\frac{(M^f)''}{M^f} = \frac{1}{\rho} \operatorname{div}_x \operatorname{div}_x \int (f - M)v \otimes v \, dv$$

$$+ \frac{v - u}{\rho T} \cdot \operatorname{div}_x \operatorname{div}_x \int (f - M)v \otimes v \otimes v \, dv$$

$$- \frac{(v - u) \cdot u}{\rho T} \operatorname{div}_x \operatorname{div}_x \int (f - M)v \otimes v \, dv$$

$$- 2 \frac{v - u}{T} \cdot \frac{\operatorname{div}_x(\rho u)}{\rho^2} \operatorname{div}_x \int (f - M)v \otimes v \, dv$$

$$+ 2 \frac{v - u}{T} \frac{u}{\rho^2} \operatorname{div}_x(\rho u) \operatorname{div}_x \int (f - M)v \, dv$$

$$+ \left(\frac{|v - u|^2}{2T} - \frac{N}{2} \right) \frac{1}{N \rho T} \operatorname{div}_x \operatorname{div}_x \int (f - M)|v|^2 v \otimes v \, dv$$

$$- \left(\frac{|v - u|^2}{2T} - \frac{N}{2} \right) \frac{1}{N \rho T} (NT + |u|^2) \operatorname{div}_x \operatorname{div}_x \int (f - M)v \otimes v \, dv$$

$$- \left(\frac{|v - u|^2}{2T} - \frac{N}{2} \right) \frac{4u}{N \rho^2 T} \operatorname{div}_x(\rho u) \left[\operatorname{div}_x \int (f - M)v \otimes v \, dv \right.$$

$$\left. - u \operatorname{div}_x \int (f - M) \, dv \right]$$

$$- \left(\frac{|v - u|^2}{2T} - \frac{N}{2} \right) \frac{2}{N \rho^2 T} \left(u \operatorname{div}_x(\rho u) - \int f v \otimes v \, dv \right)$$

$$\times \left[u \operatorname{div}_x \int (f - M) \, dv - \operatorname{div}_x \int (f - M)v \otimes v \, dv \right]$$

$$- \left(\frac{|v - u|^2}{2T} - \frac{N}{2} \right) \frac{2u}{N \rho T} \operatorname{div}_x \operatorname{div}_x \int (f - M)v \otimes v \otimes v \, dv$$

$$+ \left(\frac{|v - u|^2}{2T} - \frac{N}{2} \right) \frac{u^2}{N \rho^2 T} \operatorname{div}_x \operatorname{div}_x \int (f - M)v \otimes v \, dv$$

$$+ 4 \left(\frac{|v - u|^2}{2T} - \frac{N}{2} \right) \frac{u}{N \rho^2 T} \operatorname{div}_x(\rho u) \operatorname{div}_x \int (f - M)v \otimes v \, dv$$

$$- 4 \left(\frac{|v - u|^2}{2T} - \frac{N}{2} \right) \frac{u}{N \rho^2 T} \operatorname{div}_x(\rho u) \nabla_x \int (f - M) \, dv$$

$$- 2 \left(\frac{|v - u|^2}{2T} - \frac{N}{2} \right) \frac{\operatorname{div}_x(\rho u)}{N \rho^2 T} \operatorname{div}_x \int (f - M)|v|^2 v \, dv$$

$$+ 2 \left(\frac{|v - u|^2}{2T} - \frac{N}{2} \right) \frac{|u|^2 + NT}{N \rho^2 T} \operatorname{div}_x(\rho u) \operatorname{div}_x \int (f - M)v \, dv$$

$$
-\left(\frac{|v-u|^2}{2T}-\frac{N}{2}\right)\frac{2u}{N^2\rho^2 T}\cdot\operatorname{div}_x\int(f-M)v\otimes v\,dv
$$

$$
+\left(\frac{|v-u|^2}{2T}-\frac{N}{2}\right)\frac{2|u|^2}{N^2\rho^2 T}\operatorname{div}_x\int(f-M)v\,dv
$$

$$
-\left[-\frac{1}{\rho^2 T}\operatorname{div}_x\int fv\otimes v\,dv+\frac{u}{\rho^2 T}\operatorname{div}_x(\rho u)\right]
$$

$$
\times\left(u\operatorname{div}_x\int(f-M)\,dv-\operatorname{div}_x\int(f-M)v\otimes v\,dv\right)
$$

$$
-\frac{|v-u|^2}{2T^3}\left(-\frac{1}{\rho N}\operatorname{div}_x\int f|v|^2 v\,dv+\frac{(|u|^2+NT)}{\rho N}\operatorname{div}_x(\rho u)\right.
$$

$$
\left.+\frac{2u}{\rho N}\cdot\operatorname{div}_x\int fv\otimes v\,dv-2\frac{|u|^2}{\rho N}\nabla_x\rho\right)
$$

$$
\times\left(-\frac{1}{\rho N}\operatorname{div}_x\int(f-M)|v|^2 v\,dv+\frac{|u|^2+NT}{\rho N}\operatorname{div}_x\int(f-M)v\,dv\right.
$$

$$
\left.+\frac{2u}{\rho N}\cdot\operatorname{div}_x\int(f-M)v\otimes v\,dv-\frac{2|u|^2}{\rho N}\operatorname{div}_x\int(f-M)\,dv\right)
$$

$$
+\left(\frac{|v-u|^2}{2T}-\frac{N+4}{2}\right)\left(-\frac{1}{\rho NT}\operatorname{div}_x\int f|v|^2 v\,dv\right.
$$

$$
\left.+\left(\frac{|u|^2+NT}{\rho NT}\right)\operatorname{div}_x(\rho u)+\frac{2u}{\rho NT}\cdot\operatorname{div}_x\int fv\otimes v\,dv-2\frac{|u|^2}{\rho NT}\nabla_x\rho\right)
$$

$$
\times\frac{v-u}{\rho T}\left[u\operatorname{div}_x\int(f-M)\,dv-\operatorname{div}_x\int(f-M)v\otimes v\,dv\right]
$$

$$
+\frac{v-u}{\rho T}\left[u\operatorname{div}_x(\rho u)-\operatorname{div}_x\int fv\otimes v\,dv\right]
$$

$$
\times\frac{v-u}{\rho T}\left[u\operatorname{div}_x\int(f-M)\,dv-\operatorname{div}_x\int(f-M)v\otimes v\,dv\right]
$$

$$
-2\frac{v-u}{\rho^2 T}\cdot\left[u\operatorname{div}_x(\rho u)-\operatorname{div}_x\int fv\otimes v\,dv\right]\operatorname{div}_x\int(f-M)v\,dv
$$

$$
-\frac{2}{\rho T}\left(\frac{|v-u|^2}{2T}-\frac{N}{2}\right)\left[-\frac{1}{\rho N}\operatorname{div}_x\int f|v|^2 v\,dv+\frac{|u|^2+NT}{\rho N}\operatorname{div}_x(\rho u)\right.
$$

$$
\left.+\frac{2u}{\rho N}\cdot\operatorname{div}_x\int fv\otimes v\,dv-\frac{2|u|^2}{\rho N}\operatorname{div}_x(\rho u)\right]\operatorname{div}_x\int(f-M)v\,dv.
$$

Chapter 2
Kinetic Limits for Interacting Particle Systems

F. Rezakhanlou[*]

Abstract In this expository article, we discuss four conjectures concerning the kinetic behavior of the hard sphere models. We then formulate four stochastic variations of the hard sphere model. The known results for these stochastic models are reviewed and some proofs are sketched.

2.1 Introduction

The Boltzmann equation provides a successful description for dilute gases and can be derived from a Hamiltonian particle system. The simplest example of such a system is the hard sphere model. In the hard sphere model, one starts with N spheres of diameter ε that travel according to their velocities and collide elastically. More precisely, if $x_1, \ldots, x_N$ denote the center of the spheres and $v_1, \ldots, v_N$ denotes the velocities of the spheres, then a collision occurs between the ith and jth particle whenever $|x_i - x_j| = \varepsilon$. Their outgoing velocities after the collision are

$$v_i^j = v_i - (n_{ij} \cdot (v_i - v_j))n_{ij}, \quad v_j^i = v_j - (n_{ij} \cdot (v_i - v_j))n_{ij}, \qquad (2.1)$$

where

$$n_{ij} = \frac{x_i - x_j}{|x_i - x_j|}.$$

In a Boltzmann–Grad limit, $N \to \infty$, $\varepsilon \to 0$, but $N\varepsilon^{d-1} \to 1$. This should be compared with the fluid limit for which $N = O(\varepsilon^{-d})$. In a Boltzmann–Grad limit, we are representing a *dilute gas* because N is of a smaller order,

---

F. Rezakhanlou
Department of Mathematics, University of California Berkeley, CA 94720-3840, USA

[*]This work supported by NSF grant DMS 97–04565

namely $O(\varepsilon^{1-d})$. Note that in a dilute gas, a microscopic set of volume ε^d is empty with a high probability because on average we have $O(N\varepsilon^d) = O(\varepsilon)$ particles in such a set. For the same reason, a set of volume $O(\varepsilon^{d-1})$ has on average finitely many particles. If a particle travels for one unit of time, then it traces a set of volume ε^{d-1}. If particles scattered evenly in space, our test particle encounters $O(N\varepsilon^{d-1}) = O(1)$ many particles. In other words, the *mean free path* of a particle is positive and finite in a Boltzmann–Grad regime, and this is responsible for a *kinetic behavior*. By a "kinetic behavior" we mean various macroscopic (or mesoscopic, meaning between micro and macro) properties of the particle densities that are expected to be valid under the Boltzmann–Grad limit. By a macroscopic particle density we mean a function $f : \mathbb{R}^d \times \mathbb{R}^d \times [0,T] \to [0,\infty)$ such that

$$\varepsilon^{d-1} \sum_{i=1}^{N} J(x_i(t), v_i(t)) \; \to \; \int f(x,v,t) J(x,v) \mathrm{d}x\,\mathrm{d}v \qquad (2.2)$$

as $\varepsilon \to 0$, where J is any bounded continuous function. For example, if J is approximately the indicator function of a set A, then (2.2) says that ε^{d-1} times the number of particles in the set A is approximately $\int_A f(x,v,t)\mathrm{d}x\,\mathrm{d}v$.

Given an integrable function $f^0 : \mathbb{R}^d \times \mathbb{R}^d \times [0,\infty)$, we can distribute particles in a natural way so that the macroscopic density becomes f^0. For example, for each positive integer N, we first define ε by the formula $\varepsilon^{d-1}N = \int f^0 \mathrm{d}x\,\mathrm{d}v$. We then define the *configuration spaces* $\mathcal{E}$ to be

$$\mathcal{E} = \{\mathbf{q} = (x_1, v_1, \ldots, x_N, v_N) \in \mathbb{R}^{2Nd} : |x_i - x_j| > \varepsilon \text{ for all } i, j\}.$$

Next we define a measure μ_N on $\mathcal{E}$ by

$$\mu_N(A) = Z^{-1} \int_{A \cap \mathcal{E}} f^0(x_1, v_1) f^0(x_2, v_2) \ldots$$
$$\ldots f^0(x_N, v_N)\mathrm{d}x_1\,\mathrm{d}v_1\,\mathrm{d}x_2\,\mathrm{d}v_2 \ldots \mathrm{d}x_N\,\mathrm{d}v_N$$

with Z the normalizing constant. By standard arguments we can show that for every bounded continuous function J,

$$\lim_{N \to \infty} \int \left| \varepsilon^{d-1} \sum_{i=1}^{N} J(x_i, v_i) - \int J(x,v) f^0(x,v) \mathrm{d}x\,\mathrm{d}v \right| \mathrm{d}\mu_N = 0.$$

For each N, the probability measure μ_N induces a probability measure P_N on the space of trajectories $(\mathbf{q}(t) = (x_1(t), v_1(t), \ldots, x_N(t), v_N(t)) : t \in \mathbb{R}^+)$. The corresponding expectation is denoted by E_N.

We are now ready to formulate our first conjecture regarding the macroscopic properties of our model.

Conjecture 1. For every positive t,

$$\lim_{N \to \infty} E_N \left| \varepsilon^{d-1} \sum_{i=1}^{N} J(x_i(t), v_i(t)) - \int J(x, v) f(x, v, t) \mathrm{d}x\, \mathrm{d}v \right| = 0 ,$$

where f is the (unique) solution to the Boltzmann equation

$$f_t + v \cdot f_x = Q(f, f) = \int_{\mathbb{R}^d} \int_{\mathbb{S}} B(v - v_*, n)[f(x, v') f(x, v'_*) - f(x, v) f(x, v_*)] \mathrm{d}n\, \mathrm{d}v_*,$$
(2.3)

with the initial condition $f(x, v, t) = f^0(x, v)$. Here $\mathbb{S}$ denotes the unit sphere, $\mathrm{d}n$ denotes the $d-1$-dimensional Hausdorff measure on $\mathbb{S}$, $B(v - v_*, n) = (n \cdot (v - v_*))^+$, and

$$\begin{cases} v' = v - (n \cdot (v - v_*))n , \\ v'_* = v_* + (n \cdot (v - v_*))n . \end{cases}$$

The term $(n \cdot (v - v_*))^+ f(x, v) f(x, v_*)$ in (2.3) represents the *loss term*. Its form has to do with the celebrated *molecular chaos* assumption (or Stosszahlansatz) of Boltzmann. This assumption asserts that before a collision, particles are (approximately) stochastically independent. Hence the probability of having a pair of particles of velocities v and v_* near a point x is approximately $f(x, v) f(x, v_*)$. The *gain term* $(n \cdot (v - v_*))^+ f(x, v') f(x, v'_*)$ has a similar interpretation.

Conjecture 1 has been established under some stringent conditions. Lanford [11] proves a variant of this conjecture for "short times". (See also [10].) Illner and Pulvirenti [9], [13] replace the smallness on time with a smallness on the initial condition with respect to a suitable norm. See the book [5] for a proof.

Conjecture 1 does not tell us about the limit of

$$\varepsilon^{d-1} \sum_{i=1}^{N} J(x_i(t_1), v_i(t_1), x_i(t_2), v_i(t_2), \dots, x_i(t_k), v_i(t_k))$$
(2.4)

as $N \to \infty$ and when $k > 1$. Since the expression in (2.4) is symmetric in $(x_1, v_1, \dots, x_N, v_N)$ (in other words, relabeling particles does not change (2.4)), assume that the probability distribution P_N is also symmetric in $\mathbf{q} = (x_1, v_1, \dots, x_N, v_N)$. Now we can say that in fact (x_1, v_1) (or any other (x_i, v_i)) is a random variable with probability density almost equal to $\frac{1}{Z_0} f^0(x, v)$ with $Z_0 = \int f^0(x, v) \mathrm{d}x\, \mathrm{d}v$:

$$\lim_{N \to \infty} P_N((x_1(0), v_1(0)) \in A) = Z_0^{-1} \int_A f^0(x, v) \mathrm{d}x\, \mathrm{d}v .$$

It turns out that we can calculate the limit of

$$\varepsilon^{d-1} \sum_{i=1}^{N} J(x(\cdot), v(\cdot))$$

for every continuous functional of the trajectories $(x(s), v(s) : 0 \leq s \leq T)$ provided that we find the limiting distribution of the process $(x_1(\cdot), v_1(\cdot))$ as $N \to \infty$. In this connection we have the following conjecture:

Conjecture 2. The law of the process $(x_1(\cdot), v_1(\cdot))$ with respect to P_N converges to the law of the process $(\bar{x}(\cdot), \bar{v}(\cdot))$ as $N \to \infty$, where $(\bar{x}(\cdot), \bar{v}(\cdot))$ is an inhomogeneous Markov process with the infinitesimal generator

$$\mathcal{L}g(x,v) = v \cdot \frac{\partial g}{\partial x} + \int \int_{\mathbb{S}} [(v - v_*) \cdot n]^+ f(x, v_*, t)(g(v') - g(v)) dn\, dv_* , \quad (2.5)$$

where f solves (2.3).

A variant of this conjecture has been established by van Beijeren, Lanford, Lebowitz and Spohn [1] when the system is in an equilibrium state. (Examples of equilibrium states will be given in Sect. 2.2.)

Both Conjectures 1 and 2 do not give us any information on the size of the error we cause by replacing the microscopic density with the macroscopic density. If $f^\varepsilon(x, v, t)$ represents the microscopic density, then our Conjecture 1 asserts

$$\begin{aligned} f_t^\varepsilon + v \cdot f_x^\varepsilon &= Q(f^\varepsilon, f^\varepsilon) + o(1) , \\ f^\varepsilon &= f + o(1) , \end{aligned} \quad (2.6)$$

where $o(1)$ represents an error that goes to zero as $\varepsilon \to 0$. It turns out that $o(1)$ in (2.6) is indeed $O(1/\sqrt{N}) = O(\varepsilon^{(1/2)(d-1)})$ and this has to do with the fact that Conjecture 1 is a law of large numbers for which a central limit theorem is expected to be true. More precisely we have,

$$f_t^\varepsilon + v \cdot f_x^\varepsilon = Q(f^\varepsilon, f^\varepsilon) + \varepsilon^{(d-1)/2}\eta(x, v, t) + o(\varepsilon^{(d-1)/2}) \quad (2.7)$$

where $\eta(x, v, t)$ is a suitable space–time white noise. Since the white noise is singular and lives in a negative Sobolev space, it is more convenient to multiply η by a smooth function and integrate. To determine η (since η is Gaussian), it suffices to evaluate its variance. We expect to have the following formula for every smooth function $J(x, v, t)$:

$$\begin{aligned} E\left[\int \eta J\, dx dv\, dt\right]^2 &= \int \int_{\mathbb{S}} B(v - v_*, n) f(x, v, t) f(x, v_*, t) \\ &\quad \times [J(x, v', t) + J(x, v_*', t) - J(x, v, t) - J(x, v_*, t)]^2 \\ &\quad \times dn\, dx\, dv\, dv_*\, dt , \end{aligned} \quad (2.8)$$

where $B(v - v_*, n) = [(v - v_*) \cdot n]^+$ and E denotes the expectation.

To have a more tractable formulation of (2.7), we write

$$f^\varepsilon = f + \varepsilon^{(d-1)/2}\xi + o(\varepsilon^{(d-1)/2}) \quad (2.9)$$

and try to derive an equation for ξ. The result is:

$$\xi_t + v \cdot \xi_x = 2Q(f, \xi) + \eta. \tag{2.10}$$

Now this stochastic differential equation is linear in ξ and is simply a diffusion in an infinite dimensional setting. One may rewrite (2.10) as

$$d\xi = \mathcal{L}\xi \, dt + \mathcal{B} \, dW_t \tag{2.11}$$

where $\mathcal{L}$ is the linear operator

$$\mathcal{L}g = -v \cdot g_x + 2Q(f, g),$$

dW_t is a white noise in (x, t, v), and the linear operator $\mathcal{B}$ is defined in such a way that $\mathcal{B} \, dW_t = d\eta$. We are now ready to state our third conjecture.

Conjecture 3. Assume that the law of the process

$$\xi_\varepsilon^0(J) = \varepsilon^{-(d-1)/2} \left(\varepsilon^{d-1} \sum_{i=1}^{N} J(x_i(0), v_i(0)) - \int J(x, v) f^0(x, v) dx \, dv \right)$$

converges to the law of a process $\xi^0(J)$. Then the law of the process

$$\xi_\varepsilon(t, J) = \varepsilon^{-(d-1)/2} \left(\varepsilon^{d-1} \sum_{i=1}^{N} J(x_i(t), v_i(t)) - \int J(x, v) f(x, v, t) dx \, dv \right)$$

converges to the law of an *Ornstein–Uhlenbeck* process ξ that solves (2.10) with the initial condition ξ^0.

For simplicity, consider the case of equilibrium. Macroscopically, the equilibrium states correspond to solutions to the Boltzmann equation that are independent of x and t. (For example $f(x, v, t) = f^0(v) = (2\pi\beta)^{-d/2} \exp\left(-\frac{|v|^2}{2\beta}\right)$ for a positive constant β.) A consequence of Conjecture 3 in this case is

$$\lim_{\varepsilon \to 0} E_N \xi_\varepsilon(0, J) \xi_\varepsilon(t, \hat{J}) = \int \hat{J} T_t J \, f^0 \, dx \, dv,$$

where T_t is the semigroup generated by the linearized Boltzmann operator $\mathcal{L}$. This consequence of Conjecture 3 was established in [1] for short times. The generalization of this to nonequilibrium was carried out by Spohn [22].

A central limit theorem gives us information about the probability of small deviations. In other words, it tells us that with what probability we can have $f^\varepsilon \approx f + \varepsilon^{(d-1)/2}\xi$. For our last conjecture, we would like to formulate a problem regarding the probability of large deviations. Perhaps we should regard our microscopic density f^ε as a measure given by

$$\varepsilon^{(d-1)/2} \sum_{i=1}^{N} \delta_{(x_i(t), v_i(t))} \cdot$$

Hence $\int J f^{\varepsilon}$ is simply $\varepsilon^{(d-1)/2} \sum J(x_i(t), v_i(t))$. Conjecture 1 implies

$$\lim_{N \to \infty} P_N(f^{\varepsilon} \text{ is close to the measure } f(x, v, t) dx \, dv \text{ for } t \in [0, T]) = 1.$$

We now choose an arbitrary function g that is not the solution f and would like to know how fast the sequence

$$P_N(f^{\varepsilon} \text{ is close to the measure } g(x, v, t) dx \, dv \text{ for } t \in [0, T]) \qquad (2.12)$$

goes to zero. It is often the case that such a sequence of probabilities goes to zero exponentially fast and the exponential rate should retain some valuable information about our model. Here is our last conjecture.

Conjecture 4. Suppose

$$P_N(f^{\varepsilon} \text{ is close to the measure } g(x, v, 0) dx \, dv \text{ for } t = 0) \approx e^{-\varepsilon^{1-d} I_0(g(\cdot, 0))}$$

Then

$$P_N(f^{\varepsilon} \text{ is close to } g(x, v, t) dx \, dv \text{ for } t \in [0, T])$$
$$\approx \exp\left[-\varepsilon^{1-d}\left(I_0(g(\cdot, 0)) + \int_0^T \int L(g(x, \cdot, t), Dg(x, \cdot, t)) dx \, dt \right) \right]$$

where $Dg = g_t + v \cdot g_x$ and $L(g(\cdot), r(\cdot))$ is the convex conjugate of $H(g(\cdot), p(\cdot))$ in p-variable with

$$H(g(\cdot), p(\cdot)) = \int \int_{\mathbb{S}} B(v - v_*, n) g(v) g(v_*) \left[e^{p(v') + p(v_*') - p(v) - p(v_*)} - 1 \right] dn \, dv \, dv_*.$$

It turns out that if $\int_0^T \int L(g, Dg) dx \, dt < \infty$ then we can find a function $\hat{p}$ such that

$$g_t + v \cdot g_x = \int \int_{\mathbb{S}} B(v - v_*, n) \left[e^{\hat{p}(x,v,t) + \hat{p}(x,v_*,t) - \hat{p}(x,v',t) - \hat{p}(x,v_*',t)} g(x, v', t) \right.$$
$$\times g(x, v_*', t) - e^{\hat{p}(x,v',t) + \hat{p}(x,v_*',t) - \hat{p}(x,v,t) - \hat{p}(x,v_*,t)} g(x, v, t) g(x, v_*, t) \right] dn \, dv_*$$

and then $\int_0^T \int L(g, Dg) dx \, dt$ equals to,

$$\int_0^T \int \int_{\mathbb{S}} B(v - v_*, n) g(x, v, t) g(x, v_*, t)$$
$$\psi(\hat{p}(x, v', t) + \hat{p}(x, v_*', t) - \hat{p}(x, v, t) - \hat{p}(x, v_*, t)) dn \, dv_* \, dx \, dv \, dt,$$

where $\psi(z) = e^z(z - 1) + 1$.

So far we have formulated four conjectures for the hard sphere models and unfortunately the available results are mostly for short times. Perhaps the difficulty has to do with the fact that even the macroscopic equation is not well understood except when the dimension is one. Note however that the hard sphere model in the Boltzmann–Grad framework does not make sense when $d = 1$. Because of this, we would rather consider more general stochastic models for which $N = O(\varepsilon^{-d-\alpha})$, $\alpha \geq -1$, but now collision occurs only with small probability of order $O(\varepsilon^{\alpha+1})$ so that the mean free path is still of order one, and a kinetic behavior is still plausible. To simplify the presentation, we switch from $\mathbb{R}^d$ to the d-dimensional torus $\mathbb{T}^d$ in some of our models. Here is our first stochastic model.

Model I. The state space is $\mathcal{E} = \{\mathbf{q} = (x_1, v_1, \ldots, x_N, v_N) \in (\mathbb{T}^d \times \mathbb{R}^d)^N\}$. Each particle of location x_i travels with velocity v_i, and two particles $(x_i, v_i), (x_j, v_j)$ collide with rate $V^\varepsilon(|x_i - x_j|)B(v_i - v_j, n_{ij})$ where $V^\varepsilon(r) = \varepsilon^\alpha V(r/\varepsilon)$ with $V : [0, \infty) \to [0, \infty)$ is a continuous function of compact support with $\int V(|z|)dz = 1$ and $n_{ij} = (x_i - x_j)/|x_i - x_j|$. We assume that B is a nonnegative Lipschitz function such that $\sup B(v, n)/|v| < \infty$. More precisely, $\mathbf{q}(t)$ is a Markov process with the infinitesimal generator

$$\mathcal{A}g(\mathbf{q}) = \sum_{i=1}^{N} v_i \cdot \frac{\partial g}{\partial x_i} + \tfrac{1}{2} \sum_{i,j=1}^{N} V^\varepsilon(|x_i - x_j|)B(v_i - v_j, n_{ij})(g(S_{ij}\mathbf{q}) - g(\mathbf{q})),$$

where $S_{ij}\mathbf{q}$ is the configuration obtained from $\mathbf{q}$ by replacing (v_i, v_j) with (v_i^j, v_j^i). For this model the microscopic density is defined by

$$\int Jf^\varepsilon = \varepsilon^{d+\alpha} \sum_{i=1}^{N} J(x_i, v_i).$$

Hence the total mass of f^ε is $\varepsilon^{d+\alpha}N$ and this is assumed to be of order one. Also note that a collision occurs with a probability that is of order $O(\varepsilon^{\alpha+1})$ and with the remaining probability particles ignore each other. This is because when $V^\varepsilon(|x_i - x_j|) \neq 0$, then $|x_i - x_j| = O(\varepsilon)$ because V is of compact support. It takes a time of order $O(\varepsilon)$ for x_i and x_j to have a chance of collision. More precisely, if $A = \{(x, y) : V(\varepsilon^{-1}|x - y|) \neq 0\}$, then (x_i, x_j) has a chance of a collision only if $(x_i, x_j) \in A$. But since $B(v_i - v_j, n_{ij}) \neq 0$ implies $v_i - v_j \neq 0$, after a time of order $O(\varepsilon^{\alpha+1})$ the pair (x_i, x_j) leaves the set A and loses its chance of collision. Hence a collision happens with a probability of order $O(\varepsilon^{\alpha+1})$. Finally we remark that for Model I, the mean free path is still positive and finite. This is because in one unit of time, an ε-neighborhood of the particle's path has a volume of order $O(\varepsilon^{d-1})$ and there are $O(N\varepsilon^{d-1})$ particles in this neighborhood. But a collision occurs with probability $O(\varepsilon^{\alpha+1})$. So the total collision rate is of order $O(\varepsilon^{\alpha+1}\varepsilon^{d-1}N)$, and this is of order one.

The drawback with our Model I is that when $d = 1$ we get a trivial model because $(v_i^j, v_j^i) = (v_j, v_i)$. To avoid this, we need to violate one of the conservation laws. For this, one may replace $B(v_i - v_j, n_{ij})$ with a more flexible collision rate $K(x_i, x_j; v_i', v_j', v_i, v_j)$ where (v_i', v_j') denote the post-collision velocities. For example, we may assume the conservation of momentum so that $v'_j = v_i + v_j - v'_i$. Another possibility is that we discretize the velocity and we assume that $v_i \in \{v_\alpha : \alpha \in I\}$ where $I = \{1, 2, \ldots, k\}$.

Model II. The state space is $\mathcal{E} = \{\mathbf{q} = (x_1, \alpha_1, \ldots, x_N, \alpha_N) \in (\mathbb{R}^d \times I)^N\}$, and the generator is

$$Ag = \sum_{i=1}^N v_{\alpha_i} \cdot \frac{\partial g}{\partial x_i} + \frac{1}{2} \sum_{i,j=1}^N V^\varepsilon(|x_i - x_j|) \sum_{\gamma, \delta} K(\alpha_i \alpha_j, \gamma \delta) \left(g(S_{ij}^{\gamma \delta} \mathbf{q}) - g(\mathbf{q}) \right),$$

where $K : I \times I \times I \times I \to [0, \infty)$ and $S_{ij}^{\gamma \delta} \mathbf{q}$ is the configuration we obtain from replacing the labels of the ith and jth particles with (γ, δ). We assume some natural conditions on K:

(i) $K(\alpha\beta, \gamma\delta) = K(\beta\alpha, \gamma\delta) = K(\alpha\beta, \gamma\delta)$,

(ii) $K(\alpha\beta, \gamma\delta) = 0$ if $v_\alpha = v_\beta$,

(iii) There exists a vector $(\lambda_1, \lambda_2, \ldots, \lambda_k)$ with $\lambda_i > 0$ such that $\sum_1^k \lambda_i = 1$ and $K(\alpha\beta, \gamma\delta)\lambda_\alpha\lambda_\beta = K(\gamma\delta, \alpha\beta)\lambda_\gamma\lambda_\delta$.

(2.13)

Note that for Model II the macroscopic density can be written as $(f_\alpha(x, t) : \alpha \in I)$ and the macroscopic equation becomes

$$\frac{\partial}{\partial t} f_\alpha + v_\alpha \cdot \frac{\partial f_\alpha}{\partial x} = \sum_{\beta\gamma\delta} K(\gamma\delta, \alpha\beta) f_\gamma f_\delta - K(\alpha\beta, \gamma\delta) f_\alpha f_\beta. \quad (2.14)$$

Another possibility is that we replace the free motion by random walks. Here is our third stochastic model.

Model III. For each $\alpha \in I$, we have a probability density function $(p(z, \alpha) : z \in \mathbb{Z}^d)$ that is of finite range, irreducible of mean v_α; $\sum_z zp(z, \alpha) = v_\alpha$, $\sum_z p(z, \alpha) = 1$. Each particle $x_i \in \mathbb{Z}^d$ has a label $\alpha_i \in I$. The particle x_i jumps to site $x_i + z$ with rate $\varepsilon^{-1} p(z, \alpha_i)$, and when $x_i = x_j$, then with rate $K(\alpha_i \alpha_j, \gamma\delta)$ the particles (x_i, x_j) gain new labels (γ, δ). Hence the generator is

$$Ag(\mathbf{q}) = \varepsilon^{-1} \sum_{z,i} p(z, \alpha_i)(g(S_{z,i} \mathbf{q}) - g(\mathbf{q}))$$

$$+ \frac{1}{2} \sum_{i,j=1}^N \mathbb{1}(x_i - x_j) \sum_{\gamma, \delta} K(\alpha_i \alpha_j, \gamma\delta)(g(S_{i,j}^{\gamma,\delta} \mathbf{q}) - g(\mathbf{q})),$$

where $S_{z,i} \mathbf{q}$ is the configuration we obtain from replacing x_i with $x_i + z$.

Model IV. Our last model is a variation of Model I that is macroscopically associated with the Enskog equation. The Boltzmann Equation ceases to be valid when the density of the gas increases. In 1922, Enskog proposed a modification of the Boltzmann equation (2.3) that provided a very useful description of moderately dense gases. In the Enskog equation, the expression in the brackets in (2.3) is replaced with

$$\sigma^{d-1}\big(f(x,v')f(x-\sigma n, v'_*) - f(x,v)f(x+\sigma n, v_*)\big).\qquad(2.15)$$

See for example Resibois and De Leener [14] for more information on Enskog equation. Even though (2.5) suggests a model in which a collision occurs when particles are at distance σ, it is not known whether the Enskog equation can be derived from a Hamiltonian particle system. In spite of the original motivation behind the Enskog equation, we only regard it as a mollification of the Boltzmann equation. Technically speaking, the Enskog equation is as singular as the one dimensional Boltzmann equation Because of this, it seems plausible that some of the known techniques for one-dimensional particle systems with kinetic behavior can be utilized for particle systems associated I; the collisions are elastic and occur with probability ε^d. However, two particles experience a collision only when they are at a distance $\sigma + O(\varepsilon)$ where $\sigma > 0$ is fixed. The state space is now $\mathcal{E} = \{\mathbf{q} = (x_1, v_1, \ldots, x_N, v_N) \in (\mathbb{R}^d \times \mathbb{R}^d)^N\}$. Each particle of location x_i travels with velocity v_i, and two particles $(x_i, v_i), (x_j, v_j)$ collide with rate $V^{\sigma,\varepsilon}(|x_i - x_j|)B(v_i - v_j, -n_{ij})$ where $V^{\sigma,\varepsilon}(r) = \varepsilon^{d-1}V((r-\sigma)/\varepsilon)$ with $\hat{V} : \mathbb{R} \to [0, \infty)$ is a continuous even function of compact support with $\int \hat{V}(r)dr = 1$. We assume $B(z,n) = (z \cdot n)^+$. As before, $\mathbf{q}(t)$ is a Markov process with the infinitesimal generator

$$\mathcal{A}g(\mathbf{q}) = \sum_{i=1}^{N} v_i \cdot \frac{\partial g}{\partial x_i} + \tfrac{1}{2}\sum_{i,j=1}^{N} V^{\sigma,\varepsilon}(|x_i - x_j|)B(v_i - v_j, -n_{ij})(g(S_{ij}\mathbf{q}) - g(\mathbf{q})),$$

where $S_{ij}\mathbf{q}$ is the configuration obtained from $\mathbf{q}$ by replacing (v_i, v_j) with (v_i^j, v_j^i).

Here are some of the results available for these models.

1. In Rezakhanlou [20] Conjecture 1 was established in arbitrary dimension provided that $\alpha \geq d + 1$. In Caprino, DeMasi, Presutti, Pulvirenti [3], Conjecture 1 was established in dimension 2 for a model with four velocities of type III provided that there exists a bounded smooth solution to the macroscopic equation. In Rezakhanlou [15] the analog of Conjecture 1 was established for Model III when $d = 1$. In Caprino, Pulvirenti [4], Conjecture 1 was established for Model II with four velocities when $d = 1$ and $\alpha = 0$. Conjecture 1 for Model II in dimension one for $\alpha = 0$, was also established by Rezakhanlou and Tarver in [21]. Rezakhanlou [19] establishes Conjecture 1 for Model IV. In both [4] and [3] a cluster expansion approach of the Lanford have been utilized. The approach in [15, 19, 20] and [21] is different from [4]and [3] and takes advantage of the fact that some macroscopic tricks for the Boltzmann equation have microscopic counterparts.

2. Conjecture 2 for Model III in dimension one was established in [16].
3. Conjecture 3 for Model II with $\alpha = 0$ in arbitrary dimension was carried out in [17].
4. Conjecture 4 for Model II in dimension one for $\alpha = 0$ was established in [18].

2.2 Equilibrium Fluctuations

We devote this section to the equilibrium fluctuations. Our main goal is to sketch the proof of Conjecture 3 for Model II when $\alpha = 0$. We refer the reader to [17] for more details. As we will see later, our method of proof is not good enough to treat the hard sphere model. Throughout this section, we assume $\varepsilon = N^{-1/d}$. Recall that in Model II the infinitesimal generator is $\mathcal{A}_0 + \mathcal{A}_c$, where

$$\mathcal{A}_0 g(\mathbf{q}) = \sum_{i=1}^{N} v_{\alpha_i} \cdot \frac{\partial g}{\partial x_i}(\mathbf{q}), \tag{2.16}$$

$$\mathcal{A}_c g(\mathbf{q}) = \tfrac{1}{2} \sum_{i,j=1}^{N} V(\varepsilon^{-1}(x_i - x_j)) K(\alpha_i \alpha_j, \gamma \delta)(g(S_{ij}^{\gamma \delta} \mathbf{q}) - g(\mathbf{q})). \tag{2.17}$$

For simplicity we assume that the x-space is periodic. This means that the state space is $\mathcal{E} = (\mathbb{T}^d \times I)^N$ where $\mathbb{T}^d$ is the d-dimensional torus and $I = \{1, 2, \ldots, k\}$ is the set of labels. We define $\mathbb{T}$ to be the interval $[0,1]$ with 0 and 1 identified. In (2.17), the function $V : \mathbb{R}^d \to [0, \infty)$ is a continuous function of compact support with

$$V(z) = V(-z), \qquad \int V(z)\, dz = 1.$$

The meaning of $V(\varepsilon^{-1}(x_i - x_j))$ is as follows. The points x_i and x_j each have d coordinates in the circle $\mathbb{T}$. The rth coordinate of the difference $x_i^r - x_j^r$ is defined to be the signed distance between x_i^r and x_j^r. Hence in (2.17), $x_i - x_j$ is a point in $\mathbb{R}^d$. In the sequel, if $x_i \in \mathbb{T}^d$ and $z \in \mathbb{R}^d$, then $x_i + z$ is defined periodically and belongs to $\mathbb{T}^d$.

We now describe an invariant measure for Model II. Define $\nu(d\mathbf{q})$ by

$$\int F(\mathbf{q})\nu(d\mathbf{q}) = \int \sum_{\alpha_1 \ldots \alpha_N} F(x_1, \alpha_1, \ldots, x_N, \alpha_N) \lambda_{\alpha_1} \ldots \lambda_{\alpha_N}\, dx_1 \ldots dx_N.$$

This means that $(x_1, \alpha_1), \ldots, (x_N, \alpha_N)$ are i.i.d. variables with respect to ν, and $\nu(x_i \in A, \ \alpha_i = \alpha) = \lambda_\alpha |A|$ where $|A|$ is the Lebesgue measure of A. One can readily show that for any functions F and G,

$$\int (\mathcal{A}_0 G + \mathcal{A}_c G) H \, d\nu = \int G(-\mathcal{A}_0 H + \mathcal{A}_c H) d\nu \,.$$

In words, $\mathcal{A}_0$ is antisymmetric and $\mathcal{A}_c$ is symmetric with respect to ν. The latter follows from our assumption $K(\alpha\beta, \gamma\delta)\lambda_\alpha\lambda_\beta = K(\gamma\delta, \alpha\beta)\lambda_\gamma\lambda_\delta$.

The main step in the proof of equilibrium fluctuations is a version of *Stosszahlansatz*. Roughly, we would like to say that near a point x, the number of collisions is approximately proportional to the product of densities. To prove this, we show that shifting particles around somewhat does not dramatically alter the number of collisions. We first explain this macroscopically. Recall that our PDE is

$$f_t + v_\alpha \cdot f_x = Q_\alpha(x, t) = \sum_{\beta\gamma\delta} K(\gamma\delta, \alpha\beta) f_\gamma f_\delta - K(\alpha\beta, \gamma\delta) f_\alpha f_\beta \,.$$

We certainly have that the expression $f_\alpha(x + z, t) f_\beta(x, t)$ is equal to

$$f_\alpha^0(x + z - v_\alpha t) f_\beta^0(x - v_\beta t) + \int_0^t \frac{d}{ds} \left[f_\alpha(x + z - v_\alpha(t - s), s) f_\beta(x - v_\beta(t - s), s) \right] ds$$

$$= f_\alpha^0(x + z - v_\alpha t) f_\beta^0(x - v_\beta t) + \int_0^t Q_\alpha(x + z - v_\alpha(t-s), s) f_\beta(x - v_\beta(t-s), s) ds$$

$$+ \int_0^t f_\alpha(x + z - v_\alpha(t - s), s) Q_\beta(x - v_\beta(t - s), s) ds$$

As a result,

$$X(z) = \int_0^T \int f_\alpha(x + z, t) f_\beta(x, t) J(x) dx \, dt = \Omega_1(z) + \Omega_2(z) + \Omega_3(z) \,,$$

where

$$\Omega_1(z) = \int_0^T \int f_\alpha^0(x + z - v_\alpha t) f_\beta^0(x - v_\beta t) J(x) dx \, dt \,,$$

$$\Omega_2(z) = \int_0^T \int \int_0^t Q_\alpha(x + z - v_\alpha(t - s), s) f_\beta(x - v_\beta(t - s), s) J(x) ds \, dx \, dt$$

$$\Omega_3(z) = \int_0^T \int \int_0^t f_\alpha(x + z - v_\alpha(t - s), s) Q_\beta(x - v_\beta(t - s), s) J(x) ds \, dx \, dt \,.$$

Tartar [24, 25] uses the above calculation to obtain a bound on the collision for solutions to (2.4) in dimension one. It turns out that our version of Stosszahlansatz is the microscopic analog of the claim

$$|X(\tau(v_\alpha - v_\beta)) - X(0)| \le \text{const.} \, h(\tau) \tag{2.18}$$

for a suitable continuous function $h(\tau)$ with $h(0) = 0$. To achieve (2.18), observe that for $v = v_\alpha - v_\beta$ and $z = \tau(v_\alpha - v_\beta)$, $\tau > 0$, the expression

$\Omega_1(z) - \Omega_1(0)$ is equal to,

$$\int_0^T \int \left[f_\alpha^0(x + z + (v_\beta - v_\alpha)t) f_\beta^0(x) J(x + v_\beta t) \right.$$
$$\left. - f_\alpha^0(x + (v_\beta - v_\alpha)t) f_\beta^0(x) J(x + v_\beta t) \right] dx\, dt$$

$$= \int \int_0^T f_\beta^0(x) \left[f_\alpha^0(x + (v_\beta - v_\alpha)(t - \tau)) J(x + v_\beta t) \right.$$
$$\left. - f_\alpha^0(x + (v_\beta - v_\alpha)t) J(x + v_\beta t) \right] dx\, dt$$

$$= \int f_\beta^0(x) \int_0^T \left[f_\alpha^0(x + (v_\beta - v_\alpha)(t - \tau)) J(x + v_\beta(t - \tau)) \right.$$
$$\left. - f_\alpha^0(x + (v_\beta - v_\alpha)t) J(x + v_\beta t) dt \right] dt\, dx$$

$$+ \int_0^T f_\beta^0(x) \left[\int_0^T f_\alpha^0(x + (v_\beta - v_\alpha)(t - \tau))(J(x + v_\beta t) \right.$$
$$\left. - J(x + v_\beta(t - \tau))) dt \right] dx =: \Omega_{11} + \Omega_{12}.$$

Evidently

$$\Omega_{11} = \int f_\beta^0(x) \left[\int_{-\tau}^0 - \int_{T-\tau}^T f_\alpha^0(x + (v_\beta - v_\alpha)t) J(x + v_\beta t) dt \right] dx.$$

Now observe that if the test function J is smooth, then the term Ω_{12} is small. To show Ω_{11} is small, we need to show that $\int_a^b f^0(x + tv) dt$ is small whenever $b - a$ is small. In the same fashion we can treat $\Omega_2(z) - \Omega_2(0)$ and $\Omega_3(z) - \Omega_3(0)$. For example, the term $\Omega_3(z) - \Omega_3(0)$ is equal to

$$\int_0^T \int Q_\beta(x, s) \left[\int_0^{T-s} (f_\alpha(x - v(\theta - \tau), s) - f_\alpha(x - \theta v, s)) J(x + v_\beta \theta) d\theta \right] dx\, ds,$$

where $v = v_\alpha - v_\beta$. Again, the smallness of the expression $\Omega_3(z) - \Omega_3(0)$ follows if we show $\int_a^b f(x + \theta v, s) d\theta$ is small uniformly in s, whenever $b - a$ is small.

Recall that we are interested in

$$F(\mathbf{q}) = \frac{1}{\sqrt{N}} \sum_{i=1}^N J(x_i) \mathbb{1}(\alpha_i = \alpha)$$

where J is a smooth function with $\int J(x) dx = 0$. It is well known that

$$F(\mathbf{q}(t)) = F(\mathbf{q}(0)) + \int_0^t (\mathcal{A}_0 + \mathcal{A}_c) F(\mathbf{q}(s)) ds + M(t) \qquad (2.19)$$

where $M(\cdot)$ is a martingale. Set $\xi_\varepsilon(t, J) = F(\mathbf{q}(t))$, roughly we expect to have

$$\xi_\varepsilon(t, J) = \xi_\varepsilon(0, J) + \int_0^t \mathcal{L}\xi_\varepsilon(s, J)\mathrm{d}s + \int_0^t \mathcal{B}(J)\mathrm{d}W_s + o\left(\frac{1}{\sqrt{N}}\right). \tag{2.20}$$

It turns out that the three terms on the right-hand side of (2.19) correspond to the first three terms on the right-hand side of (2.20). The most nontrivial part of this correspondence is

$$\int_0^t \mathcal{A}_c F(\mathbf{q}(s))\mathrm{d}s = 2\int_0^t Q(\xi_\varepsilon(s, J), \lambda)\mathrm{d}s + o\left(\frac{1}{\sqrt{N}}\right), \tag{2.21}$$

where

$$Q_\alpha(f, g) = \sum_{\beta\gamma\delta} K(\gamma\delta, \alpha\beta)f_\gamma g_\delta - K(\alpha\beta, \gamma\delta)f_\alpha g_\beta.$$

This is where a version of Stosszahlansatz is needed. A straightforward calculation yields

$$\mathcal{A}_c F(\mathbf{q}) = \frac{1}{2}\sum_{i,j} V(\varepsilon^{-1}(x_i - x_j))K(\alpha_i\alpha_j, \gamma\delta)\big(J(x_i)\mathbb{1}(\gamma = \alpha) + J(x_j)\mathbb{1}(\delta = \alpha)$$
$$-J(x_i)\mathbb{1}(\alpha_i = \alpha) - J(x_j)\mathbb{1}(\alpha_j = \alpha)\big).$$

Because of this, let us study

$$\int_0^t \frac{1}{\sqrt{N}}\sum_{i,j} V(\varepsilon^{-1}(x_i(s) - x_j(s)))\mathbb{1}(\alpha_i(s) = \alpha, \alpha_j(s) = \beta)J(x_i(s))\mathrm{d}s. \tag{2.22}$$

Roughly, this corresponds to

$$\sqrt{N}\int_0^t \int f_\alpha^\varepsilon(x, s)f_\beta^\varepsilon(x, s)J(x)\mathrm{d}x\,\mathrm{d}s \tag{2.23}$$

macroscopically, where f^ε is a suitable approximation of the particle density. Since J is of zero average, we may rewrite (2.23) as

$$\sqrt{N}\int_0^t \int (f_\alpha^\varepsilon(x, s)f_\beta^\varepsilon(x, s) - \lambda_\alpha\lambda_\beta)J(x)\mathrm{d}x\,\mathrm{d}s$$
$$= \sqrt{N}\int_0^t \int [(f_\alpha^\varepsilon(x, s) - \lambda_\alpha)\lambda_\beta + (f_\beta^\varepsilon(x, s) - \lambda_\beta)\lambda_\alpha]J(x)\mathrm{d}x\,\mathrm{d}s$$
$$+\sqrt{N}\int_0^t \int (f_\alpha^\varepsilon(x, s) - \lambda_\alpha)(f_\beta^\varepsilon(x, s) - \lambda_\beta)J(x)\mathrm{d}x\,\mathrm{d}s \tag{2.24}$$
$$=: X_1(\varepsilon) + X_2(\varepsilon).$$

Note that $Q(f^\varepsilon, \lambda) + Q(\lambda, f^\varepsilon) = 2Q(f^\varepsilon, \lambda)$ is a linear combination of terms like $X_1(\varepsilon)$. The term $X_2(\varepsilon)$ is the one we want to get rid of. In fact we expect $X_2(\varepsilon)$ to be small because $\sqrt{N}(f_\alpha^\varepsilon(x, s) - \lambda_\alpha)$ is expected to be of order 1. This multiplied by $f_\beta^\varepsilon(x, s) - \lambda_\beta$ should be small.

Before taking advantage of the calculation (2.24), we need to show that (2.22) is close to (2.23). But this is our Stosszahlansatz and this is where a microscopic version of (2.18) will be needed. We are now ready to state our main lemma. This lemma also suggests what candidate for f^ε in (2.23) should be used. To ease the notation, we simply write $\mathbf{q} = (x_1, v_1, \ldots, x_N, v_N)$ for $\mathbf{q}(s)$.

Lemma 2.2.1 Let $a(\varepsilon)$ be a sequence of positive numbers such that $\lim_{\varepsilon \to 0} a(\varepsilon) = 0$. Assume that $v = v_\alpha - v_\beta \neq 0$. Then

$$\lim_{N \to \infty} \sup_{|\tau| \leq a(\varepsilon)} E^N \left[\int_0^t \frac{1}{\sqrt{N}} \sum_{i,j} \left(V(\varepsilon^{-1}(x_i - x_j + \tau v)) - V(\varepsilon^{-1}(x_i - x_j)) \right) \right.$$
$$\left. \mathbb{1}(\alpha_i = \alpha, \ \alpha_j = \beta) J(x_i) ds \right]^2 = 0.$$

Let us first explain how this lemma can be used to show that (2.22) is close to (2.23) for a suitable candidate f^ε of the particle density. Observe that Lemma 2.2.1 allows a perturbation of size $a(\varepsilon)$ in the $v = v_\alpha - v_\beta$ direction. A perturbation in the remaining directions is readily achieved if the size of perturbation is microscopically small. More precisely, let $u_1, u_2, \ldots, u_{d-1}$ be an orthonormal basis for the hyperspace orthogonal to the vector v. For any set of scalars $\tau = (\tau_1, \ldots, \tau_{d-1})$ define $u(\tau) = \tau_1 u_1 + \cdots + \tau_{d-1} u_{d-1}$. Set $\|\tau\| = \max\{|\tau_1|, \ldots, |\tau_{d-1}|\}$. The proof of the following lemma is straightforward.

Lemma 2.2.2 Let $b(\varepsilon)$ be a sequence of positive numbers such that $\lim_{\varepsilon \to 0} \frac{b(\varepsilon)}{\varepsilon} = 0$. Then

$$\lim_{\varepsilon \to 0} \sup_{\|\tau\| < b(\varepsilon)} E_N \left\{ \int_0^t \frac{1}{\sqrt{N}} \sum_{i,j} [V(\varepsilon^{-1}(x_i - x_j + u(\tau))) - V(\varepsilon^{-1}(x_i - x_j))] \right.$$
$$\left. \mathbb{1}(\alpha_i = \alpha, \ \alpha_j = \beta) J(x_i) ds \right\}^2 = 0.$$

We are now ready to define our candidate for f^ε in (2.22). Let C_ε denote a box centered at the origin with one of its axes parallel to the vector $v = v_\alpha - v_\beta$, with side length $2|v|a(\varepsilon)$ in the direction of v, and side length $2b(\varepsilon)$ in the other directions. Define $h^\varepsilon(z) = \frac{1}{Z(\varepsilon)} \mathbb{1}(z \varepsilon C_\varepsilon)$ where $Z(\varepsilon) = 2^d a(\varepsilon)|v|(b(\varepsilon))^{d-1}$ is the volume of C_ε. We then define

$$f_\alpha^\varepsilon(x,t) = \frac{1}{N}\sum_{i=1}^{N} h^\varepsilon(x_i(t)-x)\mathbb{1}(\alpha_i(t)=\alpha).$$

In fact our important Lemma 2.2.1 and straightforward Lemma 2.2.2 imply

Theorem 2.2.1 *Let C_ε as before with $\lim\limits_{\varepsilon\to 0} a(\varepsilon) = \lim\limits_{\varepsilon\to 0} b(\varepsilon)/\varepsilon = 0$, $v = v_\alpha - v_\beta \neq 0$. Then*

$$\lim_{\varepsilon\to 0} E_N\left[\int_0^t \frac{1}{\sqrt{N}}\sum_{i,j} V(\varepsilon^{-1}(x_i-x_j))\mathbb{1}(\alpha_i=\alpha,\ \alpha_j=\beta)J(x_i)ds\right.$$

$$\left.-\sqrt{N}\int_0^t\int NV(\varepsilon^{-1}(w_1-w_2))f_\alpha^\varepsilon(w_1,s)f_\beta^\varepsilon(w_2,s)J(w_1)dw_1 dw_2\,ds\right]^2 = 0.$$

To complete the proof of (2.21), we need to show that $X_2(\varepsilon) = O(1)$. For this we can drop the time integral.

Lemma 2.2.3 *Let $a(\varepsilon)$, $b(\varepsilon)$ and f^ε be as in the previous lemma. Assume $\lim\limits_{\varepsilon\to 0} a(\varepsilon)b(\varepsilon)^{d-1}N = \infty$. Then*

$$\lim_{\varepsilon\to 0}\sqrt{N}E_N\left[\int NV(\varepsilon^{-1}(w_1-w_2))(f_\alpha^\varepsilon(w_1,s)-\lambda_\alpha)\right.$$

$$\left.(f_\beta^\varepsilon(w_1,s)-\lambda_\beta)J(w_1)dw_1\,dw_2\right]^2 = 0.$$

Note that the expression on the left is independent of s and Lemma 2.2.4 is a straightforward calculation.

Before we sketch the proof of Lemma 2.2.1, let us say a few words on how Lemmas 2.2.1 and 2.2.2 imply Theorem 2.2.1. Indeed Lemmas 2.1 and 2.2 imply that

$$\int_0^t \frac{1}{\sqrt{N}}\sum_{i,j} V(\varepsilon^{-1}(x_i-x_j))\mathbb{1}(\alpha_i=\alpha,\ \alpha_j=\beta)J(x_i)ds \qquad (2.25)$$

can be replaced with

$$\int_0^t \frac{1}{\sqrt{N}}\sum_{i,j} V(\varepsilon^{-1}(x_i-x_j+u(\tau_1)-u(\tau_2)+(\theta_1-\theta_2)v))\mathbb{1}(\alpha_i=\alpha,\ \alpha_j=\beta)$$

$$\cdot\, J(x_i+u(\tau_1)+\theta_1 v)ds + O(1) \qquad (2.26)$$

provided that $\|\tau_1\|+\|\tau_2\| \le b(\varepsilon)$, $|\theta_1|+|\theta_2| \le a(\varepsilon)$ and $\lim\limits_{\varepsilon\to 0} a(\varepsilon) = \lim\limits_{\varepsilon\to 0} b(\varepsilon)/\varepsilon = 0$. As (τ_i,θ_i) varies in the set $[-b(\varepsilon),b(\varepsilon)]^{d-1}\times[-a(\varepsilon),a(\varepsilon)]$, the point $u(\tau_i)+\theta_i v$ varies in the box C_ε. We then integrate (2.26) with respect to $d\tau_1\,d\theta_1 d\tau_2 d\theta_2$ and divide the integral by "volume of $C(\varepsilon)$" $= 2^d a(\varepsilon)(b(\varepsilon))^{d-1}$.

After interchanging $d\tau_1 d\theta_1 d\tau_2 d\theta_2$ integral with the summation and making the change of variables

$$x_i + u(\tau_1) + \theta_1 v \to w_1, \qquad x_j + u(\tau_2) + \theta_2 v \to w_2$$

we deduce that (2.25) can be replaced with

$$\int_0^t \frac{1}{\sqrt{N}} \sum_{i,j} \int V(\varepsilon^{-1}(w_1 - w_2)) J(w_1) \mathbb{1}(\alpha_i(s) = \alpha) \mathbb{1}(\alpha_j(s) = \beta)$$

$$\cdot h^\varepsilon(x_i(s) - w) h^\varepsilon(x_j(s) - w_2) dw_1 \, dw_2 \, ds \qquad (2.27)$$

$$= \sqrt{N} \int_0^t \iint N V(\varepsilon^{-1}(w_1 - w_2)) J(w_1) f_\alpha^\varepsilon(w_1, s) f_\beta^\varepsilon(w_2, s) ds.$$

We now turn to the proof of Lemma 1. We would like to establish this lemma in a way similar to the derivation of (2.18) for the solutions to the Boltzmann equation. As a warm-up, let us address a simpler question. We first wonder wether

$$E_N X_\varepsilon^2 := E_N \left[\frac{1}{\sqrt{N}} \sum_{i,j} V(\varepsilon^{-1}(x_i - x_j)) \mathbb{1}(\alpha_i = \alpha, \ \alpha_j = \beta) J(x_i) \right]^2$$

is uniformly bounded in N whenever $\int J \, dx = 0$. For this, let us write $G_\varepsilon(x, y) = V(\varepsilon^{-1}(x - y)) J(x)$. Then a straightforward calculation yields

$$\int X_\varepsilon^2(\mathbf{q}) \nu(d\mathbf{q}) = (N - 1) \lambda_\alpha \lambda_\beta \int G_\varepsilon^2(x, y) dx \, dy + (N-1)(N-2) \lambda_\alpha \lambda_\beta^2$$

$$\times \int G_\varepsilon(x, y) G_\varepsilon(x, z) dx \, dy \, dz + (N-1)(N-2) \lambda_\alpha^2 \lambda_\beta$$

$$\times \int G_\varepsilon(x, y) G_\varepsilon(z, y) dx \, dy \, dz \qquad (2.28)$$

because $\int G_\varepsilon(x, y) dx \, dy = 0$. Let us write $\Omega_1(\varepsilon)$, $\Omega_2(\varepsilon)$ and $\Omega_3(\varepsilon)$ for the three terms on the right-hand side of (2.28). Evidently

$$\Omega_1(\varepsilon) = \varepsilon^d(N - 1) \lambda_\alpha \lambda_\beta \int V^2(z) dz \int J^2(z) dz$$

$$\Omega_2(\varepsilon) = \varepsilon^{2d}(N-1)(N-2) \lambda_\alpha \lambda_\beta^2 \int J^2(z) dz$$

$$\Omega_3(\varepsilon) = \varepsilon^{2d}(N-1)(N-2) \lambda_\alpha^2 \lambda_\beta \varepsilon^{-d} \int (V * V)(\varepsilon^{-1}(x - z)) J(x) J(z) dx \, dz$$

$$= \varepsilon^{2d}(N-1)(N-2) \lambda_\alpha^2 \lambda_\beta \varepsilon^{-d} \int (V * V)(\varepsilon^{-1}(x - z))(J^2(x) + O(\varepsilon)) dx \, dz$$

$$= \varepsilon^{2d}(N-1)(N-2) \lambda_\alpha^2 \lambda_\beta \int J^2(z) \, dz + O(\varepsilon),$$

where $V * V$ denotes the convolution of V with V. Hence

$$\sup_N E_N \left[\frac{1}{\sqrt{N}} \sum_{i,j} V(\varepsilon^{-1}(x_i - x_j)) \mathbb{1}(\alpha_i = \alpha, \ \alpha_j = \beta) J(x_i) \right]^2 < \infty. \quad (2.29)$$

A variation of this calculation shows that

$$\lim_{z \to 0} \sup_{\varepsilon \to 0} E_N \left[\frac{1}{\sqrt{N}} \sum_{i,j} \left(V(\varepsilon^{-1}(x_i - x_j) + z) - V(\varepsilon^{-1}(x_i - x_j)) \right) \right.$$

$$\left. \times \mathbb{1}(\alpha_i = \alpha, \ \alpha_j = \beta) J(x_i) \right]^2 = 0,$$

$$\lim_{z \to 0} \sup_{\varepsilon \to 0} E_N \left[\frac{1}{\sqrt{N}} \sum_{i,j} \left(V(\varepsilon^{-1}(x_i - x_j + z)) - V(\varepsilon^{-1}(x_i - x_j)) \right) \right.$$

$$\left. \times \mathbb{1}(\alpha_i = \alpha, \ \alpha_j = \beta) J(x_i) \right]^2 \neq 0.$$

To have equality in the latter expression, we need a time integral inside the brackets, and a vector z of the form $z = (v_\alpha - v_\beta)\tau$.

Recall that for (2.18) we used the fact that the free motion part of the dynamics has a simple semigroup and we can use the Duhamel's principle. What does this mean microscopically? To this end, let us define

$$S_t \mathbf{q} = S_t(x_1, \alpha_1, \ldots, x_N, \alpha_N) = (x_1 + v_{\alpha_1} t, \alpha_1, \ldots, x_N + v_{\alpha_N} t, \alpha_N),$$

i.e., S_t is the semigroup associated with the operator $\mathcal{A}_0$. Given a function $F(\mathbf{q})$, define $G(\mathbf{q}, \theta) = F(S_{t-\theta}\mathbf{q})$. By semigroup property,

$$E_N G(\mathbf{q}(t), t) = E_N G(\mathbf{q}(0), 0) + E_N \int_0^t (\partial_s + \mathcal{A}_0 + \mathcal{A}_c) G(\mathbf{q}(\theta), \theta) d\theta.$$

It is not hard to show that in fact $(\partial_s + \mathcal{A}_0) G(\mathbf{q}, \theta) = 0$. Hence

$$E_N F(\mathbf{q}(t)) = E_N F(S_t \mathbf{q}(0)) + E_N \int_0^t \mathcal{A}_c F(S_{t-\theta} \mathbf{q}(\theta)) d\theta. \quad (2.30)$$

To see how (2.30) can be used, put

$$R_\varepsilon(z, \mathbf{q}) = \sum_{i,j} V(\varepsilon^{-1}(x_i - x_j + z)) \mathbb{1}(\alpha_i = \alpha, \alpha_j = \beta) J(x_i),$$

$$F_\varepsilon(\mathbf{q}) = R_\varepsilon(z, \mathbf{q}) - R_\varepsilon(0, \mathbf{q}),$$

where $z = \tau v = \tau(v_\alpha - v_\beta)$. We have

$$E_N \left[\frac{1}{\sqrt{N}} \int_0^t F_\varepsilon(\mathbf{q}(s))ds \right]^2 = 2E_N \frac{1}{N} \int_0^t F_\varepsilon(\mathbf{q}(s))E_N^{\mathbf{q}(s)} \int_0^{t-s} F_\varepsilon(\mathbf{q}(\theta))d\theta ds, \tag{2.31}$$

where $E_N^{\mathbf{q}}$ denotes the expectation for the process $\mathbf{q}(\cdot)$ with $\mathbf{q}(0) = \mathbf{q}$. We now use (2.30) to assert that the expression $E_N^{\mathbf{q}} \int_0^{t-s} F_\varepsilon(\mathbf{q}(\theta))d\theta$ is equal to

$$\Omega_1(t-s, z, \mathbf{q}) + \Omega_2(t-s, z, \mathbf{q}) + \Omega_3(t-s, z, \mathbf{q}) + \Omega_4(t-s, z, \mathbf{q}) \tag{2.32}$$

where Ω_1 corresponds to the first term on the right-hand side of (2.30)

$$\Omega_1(t-s, z, \mathbf{q}) = \int_0^{t-s} \sum_{i,j} [V(\varepsilon^{-1}(x_i - x_j + z + \theta v))$$
$$-V(\varepsilon^{-1}(x_i - x_j + \theta v))]\mathbb{1}(\alpha_i = \alpha, \; \alpha_j = \beta)J(x_i + v_\alpha \theta)d\theta,$$

$$\Omega_2(t-s, z, \mathbf{q}) = E_N^{\mathbf{q}} \int_0^{t-s} \int_0^\theta \sum_{i,k} V(\varepsilon^{-1}(x_i(\sigma) - x_k(\sigma)))K(\alpha_i(\sigma)\alpha_k(\sigma), \gamma\delta)$$
$$\times \sum_j (V(\varepsilon^{-1}(x_i(\sigma) - x_j(\sigma) + v(\theta - \sigma))$$
$$-V(\varepsilon^{-1}(x_i(\sigma) - x_j(\sigma) + z + v(\theta - \sigma))))[\mathbb{1}(\gamma = \alpha) - \mathbb{1}(\alpha_i(\sigma) = \alpha)]$$
$$\times \mathbb{1}(\alpha_j(\sigma) = \beta)J(x_i(\sigma) + v_\alpha(\theta - \sigma))d\sigma\, d\theta,$$

i.e., Ω_2 corresponds to a collision between x_i with the remaining particles; Ω_3 corresponds to a collision between x_j with the remaining particles, and Ω_4 corresponds to a collision between x_i and x_j. We omit the long expressions for Ω_3 and Ω_4. Substituting (2.32) in (2.31) yields

$$E_N \left[\frac{1}{\sqrt{N}} \int_0^t F_\varepsilon(\mathbf{q}(s))ds \right]^2 = \hat{\Omega}_1 + \hat{\Omega}_2 + \hat{\Omega}_3 + \hat{\Omega}_4. \tag{2.33}$$

We would like to show that the left-hand side of (2.33) goes to zero uniformly in N as $z = (v_\alpha - v_\beta)\tau \to 0$. This is achieved by showing the same thing for each $\hat{\Omega}_r$, $r = 1, 2, 3, 4$. We first study $\hat{\Omega}_1$:

$$\hat{\Omega}_1 = \frac{2}{N}E_N \int_0^t F_\varepsilon(\mathbf{q}(s))E_N^{q(s)} \sum_{i,j} W_\varepsilon(x_i(s), x_j(s), s, \tau)\mathbb{1}(\alpha_i(s) = \alpha, \; \alpha_j(s) = \beta)$$

where

$$W_\varepsilon(x, y, s, \tau) = \int_0^{t-s} [V(\varepsilon^{-1}(x-y+\tau v+\theta v)) - V(\varepsilon^{-1}(x-y+\theta v))]J(x+v_\alpha\theta)d\theta.$$

We certainly have

$$|\hat{\Omega}_1| \le 2 \left\{ E_N \int_0^t \left[\frac{1}{\sqrt{N}} F_\varepsilon(\mathbf{q}(s)) \right]^2 ds \right\}^{1/2}$$

$$\times \left\{ E_N \int_0^t \left[\frac{1}{\sqrt{N}} \sum_{i,j} W_\varepsilon(x_i(s), x_j(s), s, \tau) \mathbb{1}(\alpha_i(s) = \alpha, \; \alpha_j(s) = \beta) \right]^2 ds \right\}^{1/2}$$

$$=: 2\Omega_{11}^{1/2} \left(\int_0^t \Omega_{12}(s) ds \right)^{1/2}$$

By stationarity and (2.29),

$$\sup_N \Omega_{11} = t \sup_N \int \left(\frac{1}{\sqrt{N}} F_\varepsilon(\mathbf{q}) \right)^2 d\nu \; < \; \infty.$$

As a result, we need to show

$$\lim_{\tau \to 0} \int_0^t \Omega_{12}(s) ds = 0.$$

We have that the term $W_\varepsilon(x, y, s, \tau)$ is equal to,

$$\int_0^{t-s} V(\varepsilon^{-1}(x - y + (\tau + \theta)v)) J(x + v_\alpha \theta) d\theta - \int_0^{t-s} V(\varepsilon^{-1}(x - y + \theta v)) J(x + v_\alpha \theta) d\theta$$

$$= \int_\tau^{t-s+\tau} V(\varepsilon^{-1}(x - y + \theta v) J(x + v_\alpha \theta) d\theta - \int_0^{t-s} V(\varepsilon^{-1}(x - y + \theta v)) J(x + v_\alpha \theta) d\theta$$

$$+ \int_0^{t-s} V(\varepsilon^{-1}(x - y + (\tau + \theta)v))(J(x + v_\alpha(\tau + \theta)) - J(x + v_\alpha \theta)) d\theta$$

$$= \int_{t-s}^{t-s+\tau} V(\varepsilon^{-1}(x - y + \theta v)) J(x + v_\alpha \theta) d\theta - \int_0^\tau V(\varepsilon^{-1}(x - y + \theta v)) J(x + v_\alpha \theta) d\theta$$

$$+ \int_0^{t-s} V(\varepsilon^{-1}(x - y + (\tau + \theta)v))(J(x + v_\alpha(\tau + \theta)) - J(x + v_\alpha \theta)) d\theta.$$

If we write W_ε^1, W_ε^2, and W_ε^3 for the three terms appeared on the right-hand side, then we have

$$W_\varepsilon(x, y, s, \tau) = W_\varepsilon^1(x, y, s, \tau) + W_\varepsilon^2(x, y, s, \tau) + W_\varepsilon^3(x, y, s, \tau).$$

With the aid of this decomposition we write

$$\Omega_{12}(s) = \Omega_{121}(s) + \Omega_{122}(s) + \Omega_{123}(s).$$

Since J is of zero average, we have

$$\iint W_\varepsilon^r(x, y, s, \tau) \mathrm{d}x \, \mathrm{d}y = 0, \quad r = 1, 2, 3.$$

Moreover,

$$|W_\varepsilon^1(x, y, s, \tau)| \le c_0 \int_{t-s}^{t-s-\tau} V(\varepsilon^{-1}(x - y + v\theta)) \mathrm{d}\theta =: \hat{W}^1(x, y)$$

$$|W_\varepsilon^2(x, y, s, \tau)| \le c_0 \int_0^\tau V(\varepsilon^{-1}(x - y + v\theta)) \mathrm{d}\theta =: \hat{W}^2(x, y)$$

$$|W_\varepsilon^3(x, y, s, \tau)| \le c_0 \tau \int_0^t V(\varepsilon^{-1}(x - y + v\theta)) \mathrm{d}\theta =: \hat{W}^3(x, y)$$

for a constant c_0. On the other hand, we can find a constant c_1 such that

$$N \iint \hat{W}_\varepsilon^r(x, y)^2 \mathrm{d}x \, \mathrm{d}y \le c_1 t^2 \tau^2$$

$$N^2 \iint \hat{W}_\varepsilon^r(x, y) \, \hat{W}_\varepsilon^r(x, z) \mathrm{d}x \, \mathrm{d}y \, \mathrm{d}z \le c_1 t^2 \tau^2$$

for $r = 1$ and 2. As in (2.29), we can put all the pieces together to deduce that

$$\left(\int_0^t \Omega_{12}(s) \mathrm{d}s \right)^{\frac{1}{2}} \le c_2 \tau$$

for some constant c_2. Hence $|\hat{\Omega}_1| \le c_3 \tau$. The treatment of $\hat{\Omega}_2$ is similar but more tedious. One can show that $\hat{\Omega}_2$ is a linear combination of terms of the form

$$\tilde{\Omega}_2 = E_N \int_0^t \frac{1}{\sqrt{N}} F_\varepsilon(\mathbf{q}(s)) E_N^{\mathbf{q}(s)} \int_0^{t-s} \frac{1}{\sqrt{N}} \sum_{i,j,k} \Lambda_\varepsilon(x_i(\sigma), x_j(\sigma), x_k(\sigma), \sigma, t)$$

$$\mathbb{1}(\alpha_i(\sigma) = \alpha', \; \alpha_j(\sigma) = \beta', \; \alpha_k(\sigma) = \gamma') \mathrm{d}\sigma \, \mathrm{d}s$$

with $v_{\alpha'} \ne v_{\gamma'}$ and the function Λ_ε that has a form similar to the function W_ε. As before, we can show that $|\tilde{\Omega}_2| \le c_4 \tau$ for some constant c_4.

Unfortunately, the above argument does not work for the hard sphere model. Hard sphere model is as singular as Model I with $\alpha = -1$. If we try to apply the above argument to Model I with $\alpha = -1$, then we run into trouble from the beginning because the bound that corresponds to (2.18) is no longer valid. This has to do with the fact that when $\alpha = 0$, then in (2.18) we are bounding the variance of a Poisson-like random variable. However, when $\alpha = -1$, the corresponding random variable is too singular to have a finite variance.

2.3 Kinetic Limits in Dimension One

Two well-known methods for the existence of solutions to a Boltzmann type equation in dimension 1 are due to Tartar [24, 25] and Bony [2]. In fact both have microscopic counterparts and can be used to establish a kinetic limit for Model II. It turns out that Bony's idea can be also used to establish a large deviations principle [18]. We already explained Tartar's techniques in Sect. 2.2.

First we explain how Bony's Lyapunov functional can be used to bound the collision. For the sake of definiteness, consider a solution to

$$f_t + v \cdot f_x = \sum_{\beta\gamma\delta} K(\gamma\delta, \alpha\beta) f_\gamma f_\delta - K(\alpha\beta, \gamma\delta) f_\alpha f_\beta \qquad (2.34)$$

Here we assume that f is periodic in x of period one. Recall

$$K(\alpha\beta, \gamma\delta) = K(\beta\alpha, \gamma\delta) = K(\alpha\beta, \gamma\delta)$$

and that $K(\alpha\beta, \gamma\delta) = 0$ if $v_\alpha = v_\beta$. We also assume the conservation of momentum:

$$K(\alpha\beta, \gamma\delta) \neq 0 \Longrightarrow v_\alpha + v_\beta = v_\gamma + v_\delta. \qquad (2.35)$$

Define the mass density

$$\rho(x, t) = \sum_\alpha f_\alpha(x, t)$$

and the momentum density,

$$u(x, t) = \sum_\alpha v_\alpha f_\alpha(x, t).$$

We then define Bony's function

$$B(t) = \int \sum_{\alpha,\beta} (v_\alpha - v_\beta)\xi(x - y)f_\alpha(x, t)f_\beta(y, t)\mathrm{d}x\,\mathrm{d}y$$

$$= \int (u(x, t)\rho(y, t) - u(y, t)\rho(x, t))\xi(x - y)\mathrm{d}x\,\mathrm{d}y$$

with $\xi(z)$ a periodic function that is defined to be $z - \frac{1}{2}$ for $z \in [0, 1)$. A straightforward calculation yields

$$\rho_t + u_x = 0$$
$$u_t + E_x = 0$$

where $E = \sum_\alpha v_\alpha^2 f_\alpha$. Hence

$$\frac{\mathrm{d}B}{\mathrm{d}t} = -\int \sum_{\alpha,\beta} (v_\alpha - v_\beta)^2 f_\alpha(x,t) f_\beta(x,t) \mathrm{d}x + \int \sum_{\alpha,\beta} (v_\alpha - v_\beta)^2 f_\alpha(x,t) f_\beta(y,t) \mathrm{d}x \, \mathrm{d}y$$

$$(2.36)$$

An immediate consequence of this is

$$\int_0^\infty \int (v_\alpha - v_\beta)^2 f_\alpha(x,t) f_\beta(x,t) \mathrm{d}x \, \mathrm{d}t \leq c_0 \left(\int \sum_\alpha f_\alpha^0(x) \mathrm{d}x \right)^2. \qquad (2.37)$$

To do this calculation microscopically, first pick $r_0 > 0$ such that $V(z) = 0$ off the interval $(-r_0, r_0)$. Then choose $W : \mathbb{R} \to \mathbb{R}$ an odd function such that $W' = V$, $W(z) = \frac{1}{2}$ for $z \geq r_0$ and $W(z) = -\frac{1}{2}$ for $z \leq -r_0$. We now need to periodize W. For this choose W^ε such that W^ε is periodic of period one and $W^\varepsilon(z) = W\left(\frac{z}{\varepsilon}\right)$ for $\frac{z}{\varepsilon} \in [-r_0, r_0]$. Such a function W^ε can be chosen so that

$$\frac{\mathrm{d}}{\mathrm{d}z} W^\varepsilon(z) = \varepsilon^{-1} V(\varepsilon^{-1} z) + R^\varepsilon(z)$$

with R^ε uniformly bounded periodic continuously differentiable, and $\frac{\mathrm{d}}{\mathrm{d}z} R^\varepsilon$ uniformly bounded. We now define

$$B^\varepsilon(\mathbf{q}) = \varepsilon^2 \sum_{i,j} W^\varepsilon(x_i - x_j)(v_{\alpha_i} - v_{\alpha_j}).$$

By semigroup property,

$$E_N B^\varepsilon(\mathbf{q}(T)) = E_N B^\varepsilon(\mathbf{q}(0)) + E_N \int_0^T (\mathcal{A}_0 + \mathcal{A}_c) B^\varepsilon(\mathbf{q}(\theta)) \mathrm{d}\theta.$$

Evidently,

$$\mathcal{A}_0 B^\varepsilon(\mathbf{q}) = \varepsilon \sum_{i,j} V(\varepsilon^{-1}(x_i - x_j))(v_{\alpha_i} - v_{\alpha_j})^2 + \varepsilon^2 \sum_{i,j} R^\varepsilon(x_i - x_j)(v_{\alpha_i} - v_{\alpha_j}).$$

It is not hard to show $\sup_N E_N B^\varepsilon(\mathbf{q}(t)) < \infty$. Also

$$\sup_N E_N \, \varepsilon^2 \sum_{i,j} R^\varepsilon(x_i - x_j)(v_{\alpha_i} - v_{\alpha_j})^2 < \infty,$$

because R^ε is uniformly bounded. To obtain

$$\sup_N E_N \int_0^T \varepsilon^{-1} \sum_{i,j} V(\varepsilon^{-1}(x_i(t) - x_j(t))) \mathbb{1}(\alpha_i(t) = \alpha, \ \alpha_j(t) = \beta) < \infty$$

whenever $v_\alpha \neq v_\beta$, it suffices to show

$$\sup_N E_N \left| \int_0^T \mathcal{A}_c B^\varepsilon(\mathbf{q}(t)) dt \right| < \infty. \tag{2.38}$$

From our macroscopic calculation (2.36), we expect that the left-hand side of (2.38) to be small. To this end, let us define

$$G^\varepsilon(y) = \varepsilon \sum_i W^\varepsilon(x_i - y)$$

so that $B^\varepsilon(\mathbf{q}) = -2\varepsilon^2 \sum_{i,j} W^\varepsilon(x_i - x_j)v_{\alpha_j} = -2\varepsilon \sum_j G(x_j)v_{\alpha_j}$. From this we deduce

$$\begin{aligned} B^\varepsilon(S_{ij}^{\gamma\delta}\mathbf{q}) - B^\varepsilon(\mathbf{q}) &= -2\varepsilon[G(x_i)(v_\gamma - v_{\alpha_i}) + G(x_j)(v_\delta - v_{\alpha_j}) \\ &= 2\varepsilon(G(x_j) - G(x_i))(v_\gamma - v_{\alpha_i}) \end{aligned}$$

whenever $v_\gamma + v_\delta = v_{\alpha_i} - v_{\alpha_j}$. As a result

$$\mathcal{A}_c B^\varepsilon(\mathbf{q}) = 2\varepsilon \sum_{i,j} K(\alpha_i\alpha_j, \gamma\delta) V(\varepsilon^{-1}(x_i - x_j))(G(x_j) - G(x_i))(v_\gamma - v_{\alpha_i}).$$

Now it is clear why $E_N|\int_0^T \mathcal{A}_c B^\varepsilon(\mathbf{q}(t)) dt|$ must be small. When $V(\varepsilon^{-1}(x_i - x_j)) \neq 0$, we have $x_i \in (x_j - \varepsilon r_0, x_j + \varepsilon r_0)$. Since $\frac{dW^\varepsilon}{dz}(z) = \varepsilon^{-1}V(\varepsilon^{-1}z) + R^\varepsilon(z)$ with $R^\varepsilon(z)$ bounded, we deduce that $|W^\varepsilon(x_k - a) - W^\varepsilon(x_k - b)| \leq c_0\varepsilon$ whenever $b \in (a - \varepsilon r_0, a + \varepsilon r_0)$ and $x_k \notin (a - 2r_0\varepsilon, a + 2r_0\varepsilon)$. As a result,

$$|G(x_i) - G(x_j)| \leq 2\varepsilon\|W^\varepsilon\|_{L^\infty} \sum_k \mathbb{1}(x_k \in (x_i - 2r_0\varepsilon, x_i + 2r_0\varepsilon)) + c_0\varepsilon N\varepsilon$$

with $c_0 = 2\max_\alpha |v_\alpha|$. Let us write $\mathcal{N}(A; \mathbf{q})$ for the total number of particles in a set A. We then have

$$|G(x_i) - G(x_j)| \leq c_1\varepsilon \sup_{|A| \leq 4r_0\varepsilon} \mathcal{N}(A; \mathbf{q}) + c_1\varepsilon$$

where the supremum is over all intervals of length at most $4r_0\varepsilon$ and c_1 is a constant. In summary, we have shown that if

$$A^\varepsilon(\mathbf{q}) = \varepsilon \sum_{i,j} V(\varepsilon^{-1}(x_i - x_j))(v_{\alpha_i} - v_{\alpha_j})^2$$

then

$$E_N \int_0^T A^\varepsilon(\mathbf{q}(t)) dt \leq c_2 + c_2 E_N \sup_{t \in [0,T]} \sup_{|A| \leq 4r_0\varepsilon} \mathcal{N}(A; \mathbf{q}(t)) \cdot \int_0^T A^\varepsilon(\mathbf{q}(t)) dt. \tag{2.39}$$

We do not expect to have $\mathcal{N}(A, \mathbf{q})$ uniformly small, but only small with a high probability. It turns out that it is more convenient to allow a random stopping time T in (2.39). In fact if we choose

$$T = \inf_{t_0} \left\{ c_2 \sup_{|A| \leq 4 r_0 \varepsilon} \mathcal{N}(A; \mathbf{q}(t)) \leq \tfrac{1}{2} \text{ for all } t < t_0 \right\}, \qquad (2.40)$$

then we have

$$E_N \int_0^T A^\varepsilon(\mathbf{q}(t)) dt \leq 2 c_2 \,.$$

But now we face the challenge of showing that

$$\lim_{N \to \infty} P_N(T > t_1) = 0 \,, \qquad (2.41)$$

for every t_1.

Note that we have two supremums in the definition of the stopping time T, one over t and the other over sets. Regarding the latter supremum, set

$$\Phi^\varepsilon(\mathbf{q}) = \varepsilon \sum_{i=1}^{N-1} \phi(\mathcal{N}([i\varepsilon, (i+1)\varepsilon); \mathbf{q}))$$

for $\phi(z) = z \log z - z + 1$. Now if

$$A \subseteq \cup_{j=r}^{r+\ell} [j\varepsilon, (j+1)\varepsilon) \,,$$

then

$$\varepsilon \mathcal{N}(A; \mathbf{q}) = \sum_{j=r}^{r+\ell} \mathcal{N}([j\varepsilon, (j+1)\varepsilon); \mathbf{q})$$

$$\leq k\ell\varepsilon + \varepsilon \sum_j \mathcal{N}([j\varepsilon, (j+1)\varepsilon); \mathbf{q}) \mathbb{1}(\mathcal{N}([j\varepsilon, (j+1)\varepsilon); \mathbf{q}) \geq k)$$

$$\leq k\ell\varepsilon + \frac{1}{\phi(k)} \Phi^\varepsilon(\mathbf{q}) \,.$$

After minimizing over k we learn that if A is an interval, then

$$\varepsilon \mathcal{N}(A; \mathbf{q}) \leq c_3 |\log(\varepsilon + |A|)|^{-1} \Phi^\varepsilon(\mathbf{q}) \,. \qquad (2.42)$$

Hence

$$T \leq \inf_{t_0} \left\{ \sup_{0 \leq t \leq t_0} \Phi^\varepsilon(\mathbf{q}(t)) \leq \frac{c}{|\log \varepsilon|} \right\}$$

for some constant c. In [21] it is shown that for some positive constants θ_0 and $c_0(t_0)$,

$$E_N \exp\left(N\theta_0 \sup_{0 \le t \le t_0} \Phi^\varepsilon(\mathbf{q}(t))\right) \le \exp(c_0(t_0)N) \qquad (2.43)$$

provided that the initial distribution of the configurations satisfy some nice conditions that macroscopically correspond to assuming $f^0 \in L^p$ for some $p > 1$. Using (2.43), it is not hard to deduce

$$\lim_{N \to \infty} N^{-1} \log P_N(T > t_1) = -\infty$$

for every positive t_1. This certainly implies what we wanted.

The bound (2.43) is the microscopic counterpart of the celebrated *entropy inequality* for the Boltzmann equation. From

$$\frac{\mathrm{d}}{\mathrm{d}t} \sum_\alpha \int f_\alpha(x,t) \log(f_\alpha(x,t)/\lambda_\alpha)\mathrm{d}x \le 0$$

we deduce

$$\sup_t \sum_\alpha \int f_\alpha(x,t) \log(f_\alpha(x,t)/\lambda_\alpha)\mathrm{d}x \le \sum_\alpha \int f_\alpha^0(x) \log(f_\alpha^0(x)/\lambda_\alpha)\mathrm{d}x .$$

Microscopically, if ν is an invariant measure as in the previous section, and if initially $\mathbf{q}$ is distributed according to $F^0(\mathbf{q})\nu(\mathrm{d}\mathbf{q})$, then at later times $\mathbf{q}(t)$ is distributed according to $F(t,\mathbf{q})\nu(\mathrm{d}\mathbf{q})$ where F solves the forward equation

$$\frac{\mathrm{d}F}{\mathrm{d}t} = (\mathcal{A}_0 + \mathcal{A}_c)^* F = (-\mathcal{A}_0 + \mathcal{A}_c)F . \qquad (2.44)$$

Here $*$ means adjoint with respect to $\mathrm{d}\nu$. From this it is not hard to deduce

$$\sup_t \int F(t,\mathbf{q}) \log F(t,\mathbf{q})\nu(\mathrm{d}\mathbf{q}) \le \int F^0(\mathbf{q}) \log F^0(\mathbf{q})\nu(\mathrm{d}\mathbf{q}) .$$

For example, if F^0 is defined by

$$\int J(\mathbf{q})F^0(\mathbf{q})\nu(\mathrm{d}\mathbf{q}) = \frac{1}{Z_0^N} \int \cdots \int \sum_{\alpha_1 \ldots \alpha_N} J(x_1, \alpha_1, \ldots, x_N, \alpha_N)$$
$$\times f_{\alpha_1}^0(x_1) \ldots f_{\alpha_N}^0(x_N)\mathrm{d}x_1 \ldots \mathrm{d}x_N$$

with $Z_0 = \int \sum_\alpha f_\alpha^0(x)\mathrm{d}x$, then $\int F^0 \log F^0 \mathrm{d}\nu$ equals $N \sum_\alpha \int (f_\alpha^0 \log \frac{f_\alpha^0}{\lambda_\alpha} - f_\alpha^0 + \lambda_\alpha)\mathrm{d}x$. Hence the assumption $\int \sum_\alpha (f_\alpha^0 \log \frac{f_\alpha^0}{\lambda_\alpha} - f_\alpha^0 + \lambda_\alpha)\mathrm{d}x < \infty$, implies,

$$\sup_t \int F(t,\mathbf{q}) \log F(t,\mathbf{q})\nu(\mathrm{d}\mathbf{q}) \le c_4 N$$

for some constant c_4. It turns out that this only implies (2.43) with the time supremum outside the expectation. To have the supremum inside, we need to assume more. Namely,

$$\int (F^0(\mathbf{q}))^p \, \nu(d\mathbf{q}) \le e^{c_5 N} \tag{2.45}$$

for some $p > 1$. This would imply

$$\int (F(t, \mathbf{q}(t)))^p \, \nu(d\mathbf{q}) \le e^{c_5 N} ,$$

and this in turn implies (2.43). In fact, if we assume F^0 is given by (2.44), then (2.45) is equivalent to assuming $f^0 \in L^p$.

Recall that an important step towards a kinetic limit is the Stosszahlansatz. As in the previous section, it suffices to prove a regularity of the collision. For example, we may try to show that the expression

$$E_N \left| \int_0^T \varepsilon \sum \left(V(\varepsilon^{-1}(x_i - x_j)) - V(\varepsilon^{-1}(x_i - x_j + z)) \right) \right.$$

$$\left. \times J(x_i) \mathbb{1}(\alpha_i = \alpha, \ \alpha_j = \beta) dt \right| , \tag{2.46}$$

is bounded above by $\gamma(|z| + \varepsilon)$, with γ a continuous function that satisfies $\gamma(0) = 0$. Macroscopically, this is done by showing

$$\left| \int_0^T \int f_\alpha(x, t)(f_\beta(x + z, t) - f_\beta(x, t)) J(x) dx dt \right| \le \gamma(|z|) , \tag{2.47}$$

whenever $v_\alpha \ne v_\beta$. This can be proved using Bony's Lyapunov functional. Consider

$$X(z, t) = \iint f_\alpha(x, t) f_\beta(x + z, t) \xi(x - y) J(x) dx \, dy .$$

Then $\frac{dX}{dt} = \sum_{j=1}^5 \Omega_j(z, t)$, where

$$\Omega_1(z, t) = -(v_\alpha - v_\beta) \int f_\alpha(x, t) f_\beta(x + z, t) J(x) dx ,$$

$$\Omega_2(z, t) = (v_\alpha - v_\beta) \int f_\alpha(x, t) f_\beta(y + z, t) J(x) dx dy ,$$

$$\Omega_3(z, t) = \int v_\alpha f_\alpha(x, t) f_\beta(y + z, t) \xi(x - y) J(x) dx dy ,$$

$$\Omega_4(z, t) = \int Q_\alpha(x, t) f_\beta(y + z, t) \xi(x - y) J(x) dx dy ,$$

$$\Omega_5(z, t) = \int f_\alpha(x, t) Q_\beta(y + z, t) \xi(x - y) J(x) dx \, dy .$$

To prove (2.47), it suffices to show

$$
\left| \int_0^T (\Omega_j(z,t) - \Omega_j(0,t)) dt \right| \le c_6 \gamma(|z|), \qquad (2.48)
$$
$$
\sup_t |X(z,t) - X(0,t)| \le c_6 \gamma(|z|)
$$

for some constant c_6. For example,

$$
\left| \int_0^T (\Omega_4(z,t) - \Omega_4(0,t)) dt \right| \le \int_0^T \int |Q_\alpha(x,t)|\,|J(x)| \int_x^{x+z} f_\beta(y,t) dy\, dx\, dt
$$
$$
\le c_7 \int_0^T \int |Q_\alpha(x,t)| dx\, dt \cdot \sup_{0 \le t \le T} \sup_x \int_x^{x+z} f_\beta(y,t) dy .
$$

As we saw before, the entropy bound implies

$$
\sup_{0 \le t \le T} \sup_x \int_x^{x+z} f_\beta(y,t) dy \le c_8 |\log|z||^{-1}
$$

and this is what we need for (2.48) when $j = 4$. This calculation can be done microscopically by looking at

$$
B^\varepsilon(\mathbf{q}(t), z) - B^\varepsilon(\mathbf{q}(t), 0)
$$

where

$$
B^\varepsilon(\mathbf{q}, z) = \varepsilon^2 \sum_{i,j} W^\varepsilon(x_i - x_j + z)(v_{\alpha_i} - v_{\alpha_j}) .
$$

We omit the details and refer to [19].

After Stosszahlansatz, what we have achieved is this: If

$$
f_\alpha^\varepsilon(x,t) = \varepsilon \sum_{i=1}^N V\left(\frac{x_i - x}{\varepsilon}\right) \mathbb{1}(\alpha_i(t) = \alpha) ,
$$

then

$$
\frac{\partial}{\partial t} f_\alpha^\varepsilon + v_\alpha \frac{\partial}{\partial x} f_\alpha^\varepsilon = Q_\alpha(f^\varepsilon * \zeta^\delta, \ f^\varepsilon * \zeta^\delta) + R^{\varepsilon,\delta} ,
$$

where $\zeta^\delta(z) = \delta^{-d} V(\frac{z}{\delta})$ and $R^{\varepsilon,\delta}$ is a stochastic error term that in small in probability as $\varepsilon \to 0$ and $\delta \to 0$, in this order. Since the functional $f \mapsto Q_\alpha(f^\varepsilon * \zeta^\delta, \ f^\varepsilon * \zeta^\delta)$ is continuous with respect to the weak topology, we can send $\varepsilon \to 0$ to achieve that the probability distributions of f^ε converge to probability distributions that are concentrated on functions f satisfying

$$
\frac{\partial f_\alpha}{\partial t} + v_\alpha \cdot \frac{\partial f_\alpha}{\partial x} = Q_\alpha(f * \zeta^\delta, \ f * \zeta^\delta) + R^\delta(f) \qquad (2.49)
$$

where $R^\delta(f) \to 0$ as $\delta \to 0$. We now need to show that $\lim_{\delta \to 0} Q_\alpha(f * \zeta^\delta, \ f * \zeta^\delta) = Q_\alpha(f, f)$. This can be done if we can establish the uniform integrality of the collision term. The proof of this is divided in two parts. First, we show that the time integral of the microscopic collision term is uniformly integrable. (This turns out to be the most technical part of the proof.) We then show that if f satisfies (2.49) and the time average of the right-hand side of (2.47) is uniformly integrable, then the macroscopic collision is uniformly integrable. See [18] and [15] for more details.

2.4 Kinetic Limits in Higher Dimensions

In this section, we sketch the proof of Conjecture 1 for Model I in the case of $\alpha = d \geq 2$. When $d \geq 2$, the best existence result available for (2.3) is due to DiPerna and Lions [6, 7]. This existence result is formulated for the so-called renormalized solutions and the uniqueness for such solutions is an open problem. Because of this the best we can hope for is that the limit points of the microscopic densities as $N \to \infty$ are all DiPerna–Lions solutions. Note however that if we already know a bounded strong solution exists, then there exists a unique renormalized solution [12].

We first explain the notion of renormalized solutions. Define

$$Q^+(f, f) = \int_{\mathbb{R}^d} \int_{SS} B(v - v_*, n) f(x, v') f(x, v'_*) dn \, dv_* ,$$
$$Q^-(f, f) = \int_{\mathbb{R}^d} \int_{SS} B(v - v_*, n) f(x, v) f(x, v_*) dn \, dv_* \tag{2.50}$$

and $Q = Q^+ - Q^-$. We say that f is a *renormalized solution* of (2.3) if

$$f \in L^1([0, \infty) \times \mathbb{T}^d \times \mathbb{R}^d), \quad f \geq 0, \quad \frac{Q^\pm(f, f)}{1 + f} \in L^1([0, T] \times \mathbb{T}^d \times \mathbb{R}^d)$$

for every positive T and for every Lipschitz continuous $\beta : [0, \infty) \to \mathbb{R}$ that satisfies

$$\sup_r (1 + r)|\beta'(r)| < \infty ,$$

we have that

$$\beta(f)_t + v \cdot \beta(f)_x = \beta'(f) Q(f, f)$$

in weak sense.

An important aspect of the Boltzmann equation is the smoothing effect of its flow term $\partial_t + v \cdot \partial_x$. This is known as the velocity averaging lemma and was quantitatively formulated and studied by Glose et al. in [8]. The velocity averaging lemma has the following flavor: If both f and $\frac{\partial f}{\partial t} + v \cdot \frac{\partial f}{\partial x}$ belong to a weakly compact subset of $L^1(\mathbb{T}^d \times \mathbb{R}^d \times [0, T])$ and ψ is a bounded smooth function, then the velocity average $\int f(x, v, t)\psi(v)dv =: \rho(x, t)$ belongs to a strongly compact subset of $L^1(\mathbb{T}^d \times \mathbb{R}^d \times [0, T])$.

Because of the nonlinearity of the collision term, a careful choice of the microscopic density will play an essential role. One possibility is to take a smooth nonnegative function $\eta : \mathbb{R}^{2d} \to \mathbb{R}$ of compact support and total integral 1, and consider

$$f^\delta(x, v; \mathbf{q}) = \varepsilon^{2d} \sum_{i=1}^N \eta^\delta (x_i - x, v_i - v) \tag{2.51}$$

where $\eta^\delta(z) = \delta^{-2d}\eta(z/\delta)$ for a small positive δ. Needless to say that for a smooth test function J,

$$\int J f^\delta \, dx \, dv = \varepsilon^{2d} \sum_{i=1}^N J(x_i, v_i) + \text{Error}(\delta)$$

where $\text{Error}(\delta) \to 0$ as $\delta \to 0$. However, when we study f^δ as a pointwise function, the behavior of f^δ depends critically on the way $\delta = \delta(\varepsilon)$ goes to zero. For example, if $\delta(\varepsilon) = \varepsilon$ then $f^\delta(x, v; \mathbf{q})$ is a *Poisson-like* random variable, and is not expected to approximate the macroscopic density for small ε. To see this, observe that whenever $\eta\left(\frac{x_i - x}{\varepsilon}, \frac{v_i - v}{\varepsilon}\right) \neq 0$, then (x_i, v_i) belongs to a set of volume $O(\varepsilon^{2d})$. If particles are scattered evenly in space, only $NO(\varepsilon^{2d}) = O(1)$ particles are involved in the calculation of $f^\varepsilon(x, v; \mathbf{q})$. As a result, we do not have enough particles to benefit from the expected ergodic property of the model. Because of this, the random function f^ε does not approximate the macroscopic particle density. In the same way, we may argue that the function $f^\varepsilon(x, v; \mathbf{q})$ is rather rough as a function of (x, v). As a crucial step in [20], we show that the velocity averages of f^ε are regular in x-variable. More precisely, if $\rho^\varepsilon(x, t) = \int f^\varepsilon(x, v; \mathbf{q}(t))\psi(v)dv$ for a smooth function ψ, then

$$E_N \sup_{|h| < \delta} \int_0^T \int |\rho^\varepsilon(x + h, t + \alpha) - \rho^\varepsilon(x, t)| dx \, dt \leq c_0 (\log\log |\log \delta|)^{-\alpha_0},$$
$$\tag{2.52}$$

where E_N denotes the expected value and $\alpha_0 = (d + 2)^{-1}(d + 3)^{-1}$. The proof of (2.52) involves an *entropy bound*, an *entropy production bound*, and the aforementioned velocity averaging lemma. In fact we can readily use the techniques of [21] to establish (2.43) for Model I, and use it to prove

$$\sup_N E_N \sup_{t \in [0,T]} \int f^\varepsilon(x, v; \mathbf{q}(t)) \log f^\varepsilon(x, v; \mathbf{q}(t)) dx \, dv < \infty. \tag{2.53}$$

Also, if $F(t, \mathbf{q})$ is as in (2.44) and

$$H(t) = \int \log F(t, \mathbf{q}) \log F(t, \mathbf{q}) \nu dq),$$

then a straightforward calculation yields

$$\frac{d}{dt}H(t) = \int (\mathcal{A} \log F)(t, \mathbf{q}) F(t, \mathbf{q}) \nu(d\mathbf{q})$$

$$= \int (\mathcal{A}_c \log F) F \, d\nu + \int \mathcal{A}_0 F \, d\nu$$

$$= \int (\mathcal{A}_c \log F) F \, d\nu + \int \mathcal{A}_c F \, d\nu.$$

We now use

$$\int (\mathcal{A}_c \log F) F d\nu = \int \sum_{i,j} V^\varepsilon(|x_i - x_j|) B(v_i - v_j, n_{ij}) \log \frac{F(S^{ij}\mathbf{q})}{F(\mathbf{q})} F(\mathbf{q}) \nu(d\mathbf{q})$$

$$= \int \sum_{i,j} V^\varepsilon(|x_i - x_j|) B(v_i - v_j, n_{ij}) \log \frac{F(\mathbf{q})}{F(S^{ij}\mathbf{q})} F(S^{ij}\mathbf{q}) \nu(d\mathbf{q}),$$

to deduce

$$\frac{d}{dt}H(t) = -\int \sum_{i,j} V^\varepsilon(|x_i - x_j|) B(v_i - v_j, n_{ij}) \psi\left(\frac{F(t, S^{ij}\mathbf{q})}{F(t, \mathbf{q})}\right) F(t, \mathbf{q}) \nu(d\mathbf{q}),$$

where $\psi(z) = z \log z - z + 1$. From this we deduce that there exists a constant c_1 such that

$$\int_0^\infty \sum_{i,j} V^\varepsilon(|x_i - x_j|) B(v_i - v_j, n_{ij}) \psi\left(\frac{F(t, S^{ij}\mathbf{q})}{F(t, \mathbf{q})}\right) F(t, \mathbf{q}) \nu(d\mathbf{q}) \leq c_1 N.$$

$$(2.54)$$

This bound is the microscopic analog of the following entropy production bound:

$$\int_0^\infty \iint_\mathbb{S} B(v - v_*, n)(f' f'_* - f f_*) \log\left(\frac{f' f'_*}{f f_*}\right) dx \, dv \, dv_* dn dt < \infty.$$

A sketch of the proof of (2.52) is in order. Using a standard stochastic calculation we can show that weakly f^ε satisfies

$$f_t^\varepsilon + v \cdot f_x^\varepsilon = \Gamma^\varepsilon + N^\varepsilon + e^\varepsilon \qquad (2.55)$$

where Γ^ε is a collision-like term, N^ε is a martingale and e^ε is an error term that goes to zero as $\varepsilon \to 0$. As in (2.50), we write $\Gamma^\varepsilon = \Gamma_+^\varepsilon - \Gamma_-^\varepsilon$ where Γ_+^ε and Γ_-^ε represent the microscopic gain and loss terms, respectively. Since no bound on the L^1-norm of $\Gamma_\pm^\varepsilon$ is available, we switch to a *renormalized* version of f^ε. The entropy bound (2.53) provides the uniform integrability of the density f^ε. This allows us to replace f^ε with $g_n^\varepsilon = \frac{f^\varepsilon}{1 + n^{-1} f^\varepsilon}$ for a large number n. It turns out that g^ε satisfies an equation similar to (2.55):

$$g_t^\varepsilon + v \cdot g_x^\varepsilon = \hat{\Gamma}^\varepsilon + \hat{N}^\varepsilon + \hat{e}^\varepsilon = \hat{\Gamma}_+^\varepsilon - \hat{\Gamma}_-^\varepsilon + \hat{N}^\varepsilon + \hat{e}^\varepsilon, \qquad (2.56)$$

where $\hat{\Gamma}_{\pm}^{\varepsilon}$ is close to a term that looks like $\Gamma_{\pm}^{\varepsilon}\left(1+n^{-1}f^{\varepsilon}\right)^{-2}$. The entropy bound (2.53) can be used to show that $\hat{\Gamma}_{-}^{\varepsilon}$ belongs to a weakly compact subset of L^1. To treat $\hat{\Gamma}_{+}^{\varepsilon}$, we use our bound on $\hat{\Gamma}_{-}^{\varepsilon}$ and the entropy production bound (2.54). As a result, the renormalized collision terms $\hat{\Gamma}_{+}^{\varepsilon}$ and $\hat{\Gamma}_{-}^{\varepsilon}$ belong to a weakly compact subset of L^1. In the same fashion, we treat the martingale term $\hat{N}^{\varepsilon}$. We then apply the velocity averaging lemma to establish (2.52).

Our main goal is the derivation of the macroscopic equation (2.3) for the particle density. For this we would like to take advantage of the regularity of ρ^{ε}. To see how this can be achieved, observe that the microscopic loss term can be expressed as

$$
\Gamma_{-}^{\varepsilon}(x,v) = \frac{1}{2}\sum_{i,j} V^{\varepsilon}(|x_i - x_j|)\eta\left(\frac{x_i - x}{\varepsilon}, \frac{v_i - v}{\varepsilon}\right) B(v_i - v_j, n_{ij})
$$

$$
= \frac{1}{2}\sum_{i}\eta\left(\frac{x_i - x}{\varepsilon}, \frac{v_i - v}{\varepsilon}\right) K(x_i, v_i),
$$

where

$$
K(x,v) = \sum_{j} V^{\varepsilon}(|x - x_j|)B\left(v - v_j, \frac{x - x_j}{|x - x_j|}\right)
$$

$$
=: \sum_{j} A\left(\frac{x - x_j}{\varepsilon}, v - v_j\right).
$$

For our rigorous derivation of the Boltzmann equation, we need to establish a suitable variant of Stosszahlansatz. Indeed our variant can be simply described in terms of the random function K. Roughly, if $\eta\left(\frac{x_i - x}{\varepsilon}, \frac{v_i - v}{\varepsilon}\right) \neq 0$ for some i, then $|x_i - x|$ and $|v_i - v|$ are of order $O(\varepsilon)$. If $K(x,v)$ is sufficiently regular in (x,v)-variables, then we can replace $K(x_i, v_i)$ with $K(x,v)$. When such a replacement is performed, we can replace $\Gamma_{-}^{\varepsilon}(x,v)$ with $f^{\varepsilon}(x,v)K(x,v)$. On the other hand

$$
K(x,v) \approx \varepsilon^{-2d}\int V^{\varepsilon}(|x - y|)B\left(v - w, \frac{x - y}{|x - y|}\right) f^{\varepsilon}(y,w)\,dy\,dw
$$

$$
\approx \iint_{\mathbb{S}} B(v - w, n)f^{\varepsilon}(x,w)\,dw\,dn.
$$

The above plausible argument explains the role of the regularity estimate (2.52) in establishing the Stosszahlansatz for the loss term. Before we move to the next step and discuss our variant of Stosszahlansatz for the gain term, let us recall that the choice of our microscopic density f^{ε} for the derivation of the macroscopic equation is *wrong*. This is because $f^{\varepsilon}(x,v)$ is a Poisson-like random object and does not approximate the macroscopic density. However, if we consider $f^{\delta(\varepsilon)}$ for a choice of $\delta(\varepsilon)$ that satisfies $\lim_{\varepsilon \to 0} \delta(\varepsilon) = 0$ and $\lim_{\varepsilon \to 0} \delta(\varepsilon)/\varepsilon = +\infty$, then $f^{\delta(\varepsilon)}$ should approximate the macroscopic density because now

$$f^{\delta(\varepsilon)}(x, v; \mathbf{q}) = \left(\frac{\varepsilon}{\delta(\varepsilon)} \right)^{2d} \sum_i \eta \left(\frac{x_i - x}{\delta(\varepsilon)}, \frac{v_i - v}{\delta(\varepsilon)} \right) \tag{2.57}$$

involves $N \times (\delta(\varepsilon))^{2d} = O \left(\left(\frac{\delta(\varepsilon)}{\varepsilon} \right)^{2d} \right)$ many particles, and since $\frac{\delta(\varepsilon)}{\varepsilon} \to \infty$, we are dealing with a large number of particles. Hence we expect $f^{\delta(\varepsilon)}$ to approximate the macroscopic density for small ε by a Law of Large Numbers. We can then derive an equation similar to (2.55) for $f^{\delta(\varepsilon)} =: \tilde{f}^\varepsilon$;

$$\tilde{f}^\varepsilon_t + v \cdot \tilde{f}^\varepsilon_x = \tilde{\Gamma}^\varepsilon + \tilde{N}^\varepsilon + \tilde{e}^\varepsilon \,,$$

where $\tilde{\Gamma}^\varepsilon$ corresponds to the collision term, $\tilde{N}^\varepsilon$ is the martingale term, and $\tilde{e}^\varepsilon$ is a small error that goes to zero as $\varepsilon \to 0$. After a renormalization, we arrive at

$$\tilde{g}^\varepsilon_t + v \cdot \tilde{g}^\varepsilon_x = \bar{\Gamma}^\varepsilon + \bar{N}^\varepsilon + \bar{e}^\varepsilon$$

where $\tilde{g}^\varepsilon = \frac{\tilde{f}^\varepsilon}{1 + n^{-1} \tilde{f}^\varepsilon}$. It turns out that $\bar{N}^\varepsilon \to 0$ as $\varepsilon \to 0$ because $\delta(\varepsilon)/\varepsilon \to +\infty$. As before, $\bar{\Gamma}^\varepsilon$ is more or less like $\tilde{\Gamma}^\varepsilon \left(1 + n^{-1} \tilde{f}^\varepsilon \right)^{-2}$. Also, $\tilde{\Gamma}^\varepsilon = \tilde{\Gamma}^\varepsilon_+ - \tilde{\Gamma}^\varepsilon_-$ where, for instance,

$$\tilde{\Gamma}^\varepsilon_-(x, v) = \frac{1}{2} \varepsilon^{2d} \sum_{i,j} V^\varepsilon(|x_i - x_j|) \eta^{\delta(\varepsilon)}(x_i - x, v_i - v) B(v_i - v_j, n_{ij})$$

$$=: \frac{1}{2} \varepsilon^{2d} \sum_i \eta^{\delta(\varepsilon)}(x_i - x, v_i - v) H(x_i, v_i) \,.$$

As before, the Stosszahlansatz can be achieved for the loss term if we can replace $H(x_i, v_i)$ with $H(x, v)$. Of course, we only need to make such a replacement for the renormalized loss term $\tilde{\Gamma}^\varepsilon_-$ that is more or less of the form $\tilde{\Gamma}^\varepsilon_- \left(1 + n^{-1} \tilde{f}^\varepsilon \right)^{-2}$. Some care is needed to carry out the replacement of $H(x_i, v_i)$ with $H(x, v)$ because H is only (x, v)-regular in L^1-sense, i.e., (2.52) holds. The renormalization factor involves $\tilde{f}^\varepsilon$ that is not so compatible with the type of expression we have for H; $H(x, v)$ is a velocity average of a density-like function that resembles f^ε and not $\tilde{f}^\varepsilon = f^{\delta(\varepsilon)}$. This creates a rather delicate situation that is handled by choosing $\delta(\varepsilon)$ in such a way that the smallness of $H(x_i, v_i) - H(x, v)$ would compensate for the incompatibility of $f^{\delta(\varepsilon)}$ with f^ε. The punchline is that we need to choose a $\delta(\varepsilon)$ that satisfies

$$\lim_{\varepsilon \to 0} \delta(\varepsilon) \varepsilon^{-1} (\log \log \log \log | \log \varepsilon |)^{-\frac{1}{2d}} = 0 \,. \tag{2.58}$$

To give a partial justification for (2.58), let us assume that something stronger that (2.52) holds for the function H, namely

$$\sup_{|a|,|w|\le\delta} |H(x+a,v+w) - H(x,v)| \le c_2 (\log\log|\log\delta|)^{-\alpha_0}, \qquad (2.59)$$

for some constant c_2. As a consequence,

$$|H(x_i,v_i) - H(x,v)| \le c_3 (\log\log|\log\varepsilon|)^{-\alpha_0},$$

for some constant c_3 and whenever $\eta^{\delta(\varepsilon)}(x_i - x, v_i - v) \ne 0$. To avoid the incompatibility of $f^{\delta(\varepsilon)}$ with f^ε, we apply the crude inequality

$$f^{\delta(\varepsilon)} \ge const. \left(\frac{\delta(\varepsilon)}{\varepsilon}\right)^{-2d} f^\varepsilon. \qquad (2.60)$$

To guarantee that the smallness of $H(x_i,v_i) - H(x,v)$ is not fully annulled by the large factor $(\delta(\varepsilon)/\varepsilon)^{2d}$, we may require

$$\lim_{\varepsilon\to 0} \delta(\varepsilon)\varepsilon^{-1}(\log\log|\log\varepsilon|)^{-(\alpha_0)/2d} = 0. \qquad (2.61)$$

It turns out that since H satisfies (2.52) instead of (2.59), the requirement on $\delta(\varepsilon)$ is (2.58) instead of (2.61).

To this end, let us assume that the function η is of the form $\eta(x,v) = \zeta(x)\zeta(v)$. Again, the gain term is approximately equal to $\tilde{\Gamma}^\varepsilon_+ \left(1 + n^{-1}\tilde{f}^\varepsilon\right)^{-2}$ where

$$\tilde{\Gamma}^\varepsilon_+(x,v) = \sum_{i,j} V^\varepsilon(|x_i - x_j|)\tilde{\zeta}^\varepsilon(x_i - x)\tilde{\zeta}^\varepsilon(v_i^j - v)B(v_i - v_j, n_{ij})$$

with $\tilde{\zeta}^\varepsilon(a) = \left(\frac{\varepsilon}{\delta(\varepsilon)}\right)^d \zeta\left(\frac{a}{\delta(\varepsilon)}\right)$. In fact the Stosszahlansatz for the gain term is achieved in two steps. In the first step, we establish a variant of Stosszahlansatz that is useful only when we show that the macroscopic density is a supersolution. This allows us to generously replace $\bar{\Gamma}^\varepsilon_+$ with a smaller quantity whenever appropriate. For example, if we define

$$u^\varepsilon(x;\mathbf{q}) = \sum_j V^\varepsilon(|x_j - x|))(|v_j| + 1),$$

and pick a smooth function J of the variable v, then we have

$$\int \bar{\Gamma}^\varepsilon_-(x,v)J(v)dv \ge \int (1 + n^{-1}\tilde{f}^\varepsilon(x,v;\mathbf{q}))^{-2} \sum_{i,j} V^\varepsilon(|x_i - x_j|)$$
$$\times \tilde{\zeta}^\varepsilon(x_i - x)\tilde{\zeta}^\varepsilon(v_i^j - v)B(v_i - v_j, n_{ij})$$
$$\times (1 + \ell^{-1}u^\varepsilon(x_i;\mathbf{q}))^{-1} J(v)\, dv$$

for every positive ℓ. We then show that the omission of the term $(1 + n^{-1}\tilde{f}^{\varepsilon}(x, v; \mathbf{q}))^{-2}$ from the right-hand side causes an error that is small for large n. This turns out to be useful because we would rather have a renormalization of the form $(1+\ell^{-1}u^{\varepsilon}(x_i; \mathbf{q}))^{-1}$ instead of $(1+n^{-1}\tilde{f}^{\varepsilon}(x, v; \mathbf{q}))^2$ when we are dealing with the gain term. This stems from the fact that $u^{\varepsilon}(x; \mathbf{q})$ is a velocity averaging for which the regularity (2.52) applies. After dropping $(1 + n^{-1}\tilde{f}^{\varepsilon}(x, v; \mathbf{q}))^{-2}$, we are left with

$$\frac{1}{2}\varepsilon^d \sum_{i,j} V^{\varepsilon}(|x_i - x_j|)B(v_i - v_j, n_{ij})\tilde{\zeta}^{\varepsilon}(x_i - x)(1+\ell^{-1}u^{\varepsilon}(x_i))^{-1}J^{\varepsilon}(v_i^j) \quad (2.62)$$

where $J^{\varepsilon}(v) = \varepsilon^{-d}\int \tilde{\zeta}^{\varepsilon}(v - w)J(w)dw$. We can now express (2.46) as

$$\sum_i \tilde{\zeta}^{\varepsilon}(x_i - x)\tilde{H}^{\varepsilon}(x_i, v_i)(1 + \ell^{-1}u^{\varepsilon}(x_i))^{-1}$$

where

$$\tilde{H}^{\varepsilon}(x, v) = \frac{1}{2}\sum_j V^{\varepsilon}(|x - x_j|)B\left(v - v_j, \frac{x - x_j}{|x - x_j|}\right)$$
$$\times J^{\varepsilon}\left(v - (v - v_j)\frac{x - x_j}{|x - x_j|}\frac{x - x_j}{|x - x_j|}\right).$$

It turns out that now we are in a position to repeat our treatment for the loss term where $\tilde{H}^{\varepsilon}(1 + \ell^{-1}u^{\varepsilon})^{-1}$ plays the role of H^{ε}.

A variant of Stosszahlansatz will be needed when we treat the macroscopic densities as subsolutions. This time we study

$$\left(1 + n^{-1}\bar{u}^{\varepsilon}(x)\right)\int \tilde{\Gamma}_+^{\varepsilon}(x, v)J(v)dv$$

where

$$\tilde{u}^{\varepsilon}(x) = \varepsilon^d \sum_j \tilde{\zeta}^{\varepsilon}(x_j - x)(|v_j|^{3/2} + 1).$$

Our Stosszahlansatz for the collision term allows us to replace the microscopic collisions with suitable nonlinear functional of densities that enjoys some stabilities with respect to the weak topology. We refer the reader to [20] for details.

Acknowledgements Most of this article is based on a minicourse that was given by the author at Centre Emile Borel of Institut Henri Poincaré during his visit in Fall 2000. The author thank Francois Golse and Stefano Olla for invitation and CNRS for a generous financial support.

References

1. H. van Beijeren, O. E. Lanford, J. L. Lebowitz and H. Spohn, *Equilibrium time correlation functions in the low-density limit.* J. Statist. Phys. **22** (1980), 237–257.
2. J-M. Bony, *Solutions globales bornées pour les modèles discrets de l'équation de Boltzmann en dimension 1 d'espace.* In Journées "Equations aux Derivées Partielles" Saint-Jean-de-Monts, 1987.
3. S. Caprino, A. De Masi, E. Presutti, M. Pulvirenti, *A derivation of the Broadwell equation.* Comm. Math. Phys. **135** (1991), 443–465.
4. S. Caprino and M. Pulvirenti, *A cluster expansion approach to a one-dimensional Boltzmann equation: a validity result.* Comm. Math. Phys. **166** (1995), 603–631.
5. C. Cercignani, R. Illner and M. Pulvirenti, "The Mathematical Theory of Dilute Gases", Springer–Verlag, 1994.
6. R. J. DiPerna and P. L. Lions, *On the Cauchy problem for Boltzmann equations: Global existence and weak stability,* Annals of Math. **130** (1989), 321–366.
7. R. J. DiPerna and P. L. Lions, *Global solutions of Boltzmann equation and the entropy inequality,* Arch. Rational Mech. Anal. **114** (1991), 47–55.
8. F. Golse, P. L. Lions, B. Perthame and R. Sentis, *Regularity of the moments of the solution of a transport equation,* J. Func. Anal. **76** (1988), 110–125.
9. R. Illner and M. Pulvirenti, *Global validity of the Boltzmann equation for a two and three-dimensional rare gas in vacuum,* Comm. Math. Phys. **105** (1986), 189–203. *Erratum and improved results,* Comm. Math. Phys. **121** (1989), 143–146.
10. F. King, *BBGKY hierarchy for positive potentials,* Ph.D. Thesis, Dept. of Mathematics, University of California at Berkeley, (1975).
11. O. E. Lanford (III), *Time evolution of large classical systems,* in "Lecture Notes in Physics", editor J. Moser, vol. 38, Springer–Verlag, Berlin, 1975.
12. P. L. Lions, *Compactness in Boltzmann's equation via Fourier integral operators and applications I, II, III,* J. Math. Kyoto Univ. **34** (1994), 391–427, 429–461, 539–584.
13. M. Pulvirenti, *Global validity of the Boltzmann equation for a three-dimensional rare gas in vacuum,* Comm. Math. Phys. **113** (1987), 79–85.
14. P. Resibois and M. De Leener, "Classical Kinetic Theory of Fluids". Wiley, New York 1977.
15. F. Rezakhanlou, *Kinetic limits for a class of interacting particle systems,* Probab. Theory Related Fields, **104** (1996), 97–146.
16. F. Rezakhanlou, *Propagation of chaos for particle systems associated with discrete Boltzmann equation,* Stochastic Processes and their Applications, **64** (1996), 55–72.
17. F. Rezakhanlou, *Equilibrium fluctuations for the discrete Boltzmann equation,* Duke Journal of Mathematics, **93** (1998), 257–288.
18. F. Rezakhanlou, *Large deviations from a kinetic limit,* Annals of Probability, **26** (1998), 1259–1340.
19. F. Rezakhanlou, *A stochastic model associated with Enskog equation and its kinetic limit,* Comm. Math. Phys. **232** (2003), 327–375.
20. F. Rezakhanlou, *Boltzmann–Grad limits for stochastic hard sphere model,* Comm. Math. Phys. **248** (2004), 553–637.
21. F. Rezakhanlou and J. L. Tarver (III), *Boltzmann–Grad limit for a particle system in continuum,* Ann. Inst. Henri Poincaré, **33** (1997), 753–796.
22. H. Spohn, "Large Scale Dynamics of Interacting Particles", Springer–Verlag, 1991.
23. H. Spohn, *Fluctuations around the Boltzmann equation,* J. Statist. Phys. **26** (1981), 285–305.
24. L. Tartar, *Existence globale pour un système hyperbolique semilinéaire de la théorie cinétique des gaz,* Séminaire Goulaouic-Schwartz (1975/1976), 1976.
25. L. Tartar, *Some existence theorems for semilinear hyperbolic systems in one space variable,* Technical report, University of Wisconsin-Madison, 1980.

Index

Lecture Notes in Mathematics

For information about earlier volumes
please contact your bookseller or Springer
LNM Online archive: springerlink.com

Vol. 1781: E. Bolthausen, E. Perkins, A. van der Vaart, Lectures on Probability Theory and Statistics. Ecole d' Eté de Probabilités de Saint-Flour XXIX-1999. Editor: P. Bernard (2002)

Vol. 1782: C.-H. Chu, A. T.-M. Lau, Harmonic Functions on Groups and Fourier Algebras (2002)

Vol. 1783: L. Grüne, Asymptotic Behavior of Dynamical and Control Systems under Perturbation and Discretization (2002)

Vol. 1784: L. H. Eliasson, S. B. Kuksin, S. Marmi, J.-C. Yoccoz, Dynamical Systems and Small Divisors. Cetraro, Italy 1998. Editors: S. Marmi, J.-C. Yoccoz (2002)

Vol. 1785: J. Arias de Reyna, Pointwise Convergence of Fourier Series (2002)

Vol. 1786: S. D. Cutkosky, Monomialization of Morphisms from 3-Folds to Surfaces (2002)

Vol. 1787: S. Caenepeel, G. Militaru, S. Zhu, Frobenius and Separable Functors for Generalized Module Categories and Nonlinear Equations (2002)

Vol. 1788: A. Vasil'ev, Moduli of Families of Curves for Conformal and Quasiconformal Mappings (2002)

Vol. 1789: Y. Sommerhäuser, Yetter-Drinfel'd Hopf algebras over groups of prime order (2002)

Vol. 1790: X. Zhan, Matrix Inequalities (2002)

Vol. 1791: M. Knebusch, D. Zhang, Manis Valuations and Prüfer Extensions I: A new Chapter in Commutative Algebra (2002)

Vol. 1792: D. D. Ang, R. Gorenflo, V. K. Le, D. D. Trong, Moment Theory and Some Inverse Problems in Potential Theory and Heat Conduction (2002)

Vol. 1793: J. Cortés Monforte, Geometric, Control and Numerical Aspects of Nonholonomic Systems (2002)

Vol. 1794: N. Pytheas Fogg, Substitution in Dynamics, Arithmetics and Combinatorics. Editors: V. Berthé, S. Ferenczi, C. Mauduit, A. Siegel (2002)

Vol. 1795: H. Li, Filtered-Graded Transfer in Using Noncommutative Gröbner Bases (2002)

Vol. 1796: J.M. Melenk, hp-Finite Element Methods for Singular Perturbations (2002)

Vol. 1797: B. Schmidt, Characters and Cyclotomic Fields in Finite Geometry (2002)

Vol. 1798: W.M. Oliva, Geometric Mechanics (2002)

Vol. 1799: H. Pajot, Analytic Capacity, Rectifiability, Menger Curvature and the Cauchy Integral (2002)

Vol. 1800: O. Gabber, L. Ramero, Almost Ring Theory (2003)

Vol. 1801: J. Azéma, M. Émery, M. Ledoux, M. Yor (Eds.), Séminaire de Probabilités XXXVI (2003)

Vol. 1802: V. Capasso, E. Merzbach, B. G. Ivanoff, M. Dozzi, R. Dalang, T. Mountford, Topics in Spatial Stochastic Processes. Martina Franca, Italy 2001. Editor: E. Merzbach (2003)

Vol. 1803: G. Dolzmann, Variational Methods for Crystalline Microstructure – Analysis and Computation (2003)

Vol. 1804: I. Cherednik, Ya. Markov, R. Howe, G. Lusztig, Iwahori-Hecke Algebras and their Representation Theory. Martina Franca, Italy 1999. Editors: V. Baldoni, D. Barbasch (2003)

Vol. 1805: F. Cao, Geometric Curve Evolution and Image Processing (2003)

Vol. 1806: H. Broer, I. Hoveijn. G. Lunther, G. Vegter, Bifurcations in Hamiltonian Systems. Computing Singularities by Gröbner Bases (2003)

Vol. 1807: V. D. Milman, G. Schechtman (Eds.), Geometric Aspects of Functional Analysis. Israel Seminar 2000-2002 (2003)

Vol. 1808: W. Schindler, Measures with Symmetry Properties (2003)

Vol. 1809: O. Steinbach, Stability Estimates for Hybrid Coupled Domain Decomposition Methods (2003)

Vol. 1810: J. Wengenroth, Derived Functors in Functional Analysis (2003)

Vol. 1811: J. Stevens, Deformations of Singularities (2003)

Vol. 1812: L. Ambrosio, K. Deckelnick, G. Dziuk, M. Mimura, V. A. Solonnikov, H. M. Soner, Mathematical Aspects of Evolving Interfaces. Madeira, Funchal, Portugal 2000. Editors: P. Colli, J. F. Rodrigues (2003)

Vol. 1813: L. Ambrosio, L. A. Caffarelli, Y. Brenier, G. Buttazzo, C. Villani, Optimal Transportation and its Applications. Martina Franca, Italy 2001. Editors: L. A. Caffarelli, S. Salsa (2003)

Vol. 1814: P. Bank, F. Baudoin, H. Föllmer, L.C.G. Rogers, M. Soner, N. Touzi, Paris-Princeton Lectures on Mathematical Finance 2002 (2003)

Vol. 1815: A. M. Vershik (Ed.), Asymptotic Combinatorics with Applications to Mathematical Physics. St. Petersburg, Russia 2001 (2003)

Vol. 1816: S. Albeverio, W. Schachermayer, M. Talagrand, Lectures on Probability Theory and Statistics. Ecole d'Eté de Probabilités de Saint-Flour XXX-2000. Editor: P. Bernard (2003)

Vol. 1817: E. Koelink, W. Van Assche (Eds.), Orthogonal Polynomials and Special Functions. Leuven 2002 (2003)

Vol. 1818: M. Bildhauer, Convex Variational Problems with Linear, nearly Linear and/or Anisotropic Growth Conditions (2003)

Vol. 1819: D. Masser, Yu. V. Nesterenko, H. P. Schlickewei, W. M. Schmidt, M. Waldschmidt, Diophantine Approximation. Cetraro, Italy 2000. Editors: F. Amoroso, U. Zannier (2003)

Vol. 1820: F. Hiai, H. Kosaki, Means of Hilbert Space Operators (2003)

Vol. 1821: S. Teufel, Adiabatic Perturbation Theory in Quantum Dynamics (2003)

Vol. 1822: S.-N. Chow, R. Conti, R. Johnson, J. Mallet-Paret, R. Nussbaum, Dynamical Systems. Cetraro, Italy 2000. Editors: J. W. Macki, P. Zecca (2003)

Vol. 1823: A. M. Anile, W. Allegretto, C. Ringhofer, Mathematical Problems in Semiconductor Physics. Cetraro, Italy 1998. Editor: A. M. Anile (2003)

Vol. 1824: J. A. Navarro González, J. B. Sancho de Salas, $\mathscr{C}^\infty$ – Differentiable Spaces (2003)

Vol. 1825: J. H. Bramble, A. Cohen, W. Dahmen, Multiscale Problems and Methods in Numerical Simulations, Martina Franca, Italy 2001. Editor: C. Canuto (2003)

Vol. 1826: K. Dohmen, Improved Bonferroni Inequalities via Abstract Tubes. Inequalities and Identities of Inclusion-Exclusion Type. VIII, 113 p, 2003.

Vol. 1827: K. M. Pilgrim, Combinations of Complex Dynamical Systems. IX, 118 p, 2003.

Vol. 1828: D. J. Green, Gröbner Bases and the Computation of Group Cohomology. XII, 138 p, 2003.

Vol. 1829: E. Altman, B. Gaujal, A. Hordijk, Discrete-Event Control of Stochastic Networks: Multimodularity and Regularity. XIV, 313 p, 2003.

Vol. 1830: M. I. Gil', Operator Functions and Localization of Spectra. XIV, 256 p, 2003.

Vol. 1831: A. Connes, J. Cuntz, E. Guentner, N. Higson, J. E. Kaminker, Noncommutative Geometry, Martina Franca, Italy 2002. Editors: S. Doplicher, L. Longo (2004)

Vol. 1832: J. Azéma, M. Émery, M. Ledoux, M. Yor (Eds.), Séminaire de Probabilités XXXVII (2003)

Vol. 1833: D.-Q. Jiang, M. Qian, M.-P. Qian, Mathematical Theory of Nonequilibrium Steady States. On the Frontier of Probability and Dynamical Systems. IX, 280 p, 2004.

Vol. 1834: Yo. Yomdin, G. Comte, Tame Geometry with Application in Smooth Analysis. VIII, 186 p, 2004.

Vol. 1835: O.T. Izhboldin, B. Kahn, N.A. Karpenko, A. Vishik, Geometric Methods in the Algebraic Theory of Quadratic Forms. Summer School, Lens, 2000. Editor: J.-P. Tignol (2004)

Vol. 1836: C. Năstăsescu, F. Van Oystaeyen, Methods of Graded Rings. XIII, 304 p, 2004.

Vol. 1837: S. Tavaré, O. Zeitouni, Lectures on Probability Theory and Statistics. Ecole d'Eté de Probabilités de Saint-Flour XXXI-2001. Editor: J. Picard (2004)

Vol. 1838: A.J. Ganesh, N.W. O'Connell, D.J. Wischik, Big Queues. XII, 254 p, 2004.

Vol. 1839: R. Gohm, Noncommutative Stationary Processes. VIII, 170 p, 2004.

Vol. 1840: B. Tsirelson, W. Werner, Lectures on Probability Theory and Statistics. Ecole d'Eté de Probabilités de Saint-Flour XXXII-2002. Editor: J. Picard (2004)

Vol. 1841: W. Reichel, Uniqueness Theorems for Variational Problems by the Method of Transformation Groups (2004)

Vol. 1842: T. Johnsen, A. L. Knutsen, K_3 Projective Models in Scrolls (2004)

Vol. 1843: B. Jefferies, Spectral Properties of Noncommuting Operators (2004)

Vol. 1844: K.F. Siburg, The Principle of Least Action in Geometry and Dynamics (2004)

Vol. 1845: Min Ho Lee, Mixed Automorphic Forms, Torus Bundles, and Jacobi Forms (2004)

Vol. 1846: H. Ammari, H. Kang, Reconstruction of Small Inhomogeneities from Boundary Measurements (2004)

Vol. 1847: T.R. Bielecki, T. Björk, M. Jeanblanc, M. Rutkowski, J.A. Scheinkman, W. Xiong, Paris-Princeton Lectures on Mathematical Finance 2003 (2004)

Vol. 1848: M. Abate, J. E. Fornaess, X. Huang, J. P. Rosay, A. Tumanov, Real Methods in Complex and CR Geometry, Martina Franca, Italy 2002. Editors: D. Zaitsev, G. Zampieri (2004)

Vol. 1849: Martin L. Brown, Heegner Modules and Elliptic Curves (2004)

Vol. 1850: V. D. Milman, G. Schechtman (Eds.), Geometric Aspects of Functional Analysis. Israel Seminar 2002-2003 (2004)

Vol. 1851: O. Catoni, Statistical Learning Theory and Stochastic Optimization (2004)

Vol. 1852: A.S. Kechris, B.D. Miller, Topics in Orbit Equivalence (2004)

Vol. 1853: Ch. Favre, M. Jonsson, The Valuative Tree (2004)

Vol. 1854: O. Saeki, Topology of Singular Fibers of Differential Maps (2004)

Vol. 1855: G. Da Prato, P.C. Kunstmann, I. Lasiecka, A. Lunardi, R. Schnaubelt, L. Weis, Functional Analytic Methods for Evolution Equations. Editors: M. Iannelli, R. Nagel, S. Piazzera (2004)

Vol. 1856: K. Back, T.R. Bielecki, C. Hipp, S. Peng, W. Schachermayer, Stochastic Methods in Finance, Bressanone/Brixen, Italy, 2003. Editors: M. Fritelli, W. Runggaldier (2004)

Vol. 1857: M. Émery, M. Ledoux, M. Yor (Eds.), Séminaire de Probabilités XXXVIII (2005)

Vol. 1858: A.S. Cherny, H.-J. Engelbert, Singular Stochastic Differential Equations (2005)

Vol. 1859: E. Letellier, Fourier Transforms of Invariant Functions on Finite Reductive Lie Algebras (2005)

Vol. 1860: A. Borisyuk, G.B. Ermentrout, A. Friedman, D. Terman, Tutorials in Mathematical Biosciences I. Mathematical Neuroscience (2005)

Vol. 1861: G. Benettin, J. Henrard, S. Kuksin, Hamiltonian Dynamics – Theory and Applications, Cetraro, Italy, 1999. Editor: A. Giorgilli (2005)

Vol. 1862: B. Helffer, F. Nier, Hypoelliptic Estimates and Spectral Theory for Fokker-Planck Operators and Witten Laplacians (2005)

Vol. 1863: H. Führ, Abstract Harmonic Analysis of Continuous Wavelet Transforms (2005)

Vol. 1864: K. Efstathiou, Metamorphoses of Hamiltonian Systems with Symmetries (2005)

Vol. 1865: D. Applebaum, B.V. R. Bhat, J. Kustermans, J. M. Lindsay, Quantum Independent Increment Processes I. From Classical Probability to Quantum Stochastic Calculus. Editors: M. Schürmann, U. Franz (2005)

Vol. 1866: O.E. Barndorff-Nielsen, U. Franz, R. Gohm, B. Kümmerer, S. Thorbjønsen, Quantum Independent Increment Processes II. Structure of Quantum Lévy Processes, Classical Probability, and Physics. Editors: M. Schürmann, U. Franz, (2005)

Vol. 1867: J. Sneyd (Ed.), Tutorials in Mathematical Biosciences II. Mathematical Modeling of Calcium Dynamics and Signal Transduction. (2005)

Vol. 1868: J. Jorgenson, S. Lang, $Pos_n(R)$ and Eisenstein Series. (2005)

Vol. 1869: A. Dembo, T. Funaki, Lectures on Probability Theory and Statistics. Ecole d'Eté de Probabilités de Saint-Flour XXXIII-2003. Editor: J. Picard (2005)

Vol. 1870: V.I. Gurariy, W. Lusky, Geometry of Müntz Spaces and Related Questions. (2005)

Vol. 1871: P. Constantin, G. Gallavotti, A.V. Kazhikhov, Y. Meyer, S. Ukai, Mathematical Foundation of Turbulent Viscous Flows, Martina Franca, Italy, 2003. Editors: M. Cannone, T. Miyakawa (2006)

Vol. 1872: A. Friedman (Ed.), Tutorials in Mathematical Biosciences III. Cell Cycle, Proliferation, and Cancer (2006)

Vol. 1873: R. Mansuy, M. Yor, Random Times and Enlargements of Filtrations in a Brownian Setting (2006)

Vol. 1874: M. Yor, M. Émery (Eds.), In Memoriam Paul-André Meyer - Séminaire de Probabilités XXXIX (2006)

Vol. 1875: J. Pitman, Combinatorial Stochastic Processes. Ecole d'Eté de Probabilités de Saint-Flour XXXII-2002. Editor: J. Picard (2006)

Vol. 1876: H. Herrlich, Axiom of Choice (2006)

Vol. 1877: J. Steuding, Value Distributions of L-Functions (2007)

Vol. 1878: R. Cerf, The Wulff Crystal in Ising and Percolation Models, Ecole d'Eté de Probabilités de Saint-Flour XXXIV-2004. Editor: Jean Picard (2006)

Vol. 1879: G. Slade, The Lace Expansion and its Applications, Ecole d'Eté de Probabilités de Saint-Flour XXXIV-2004. Editor: Jean Picard (2006)

Vol. 1880: S. Attal, A. Joye, C.-A. Pillet, Open Quantum Systems I, The Hamiltonian Approach (2006)

Vol. 1881: S. Attal, A. Joye, C.-A. Pillet, Open Quantum Systems II, The Markovian Approach (2006)

Vol. 1882: S. Attal, A. Joye, C.-A. Pillet, Open Quantum Systems III, Recent Developments (2006)

Vol. 1883: W. Van Assche, F. Marcellàn (Eds.), Orthogonal Polynomials and Special Functions, Computation and Application (2006)

Vol. 1884: N. Hayashi, E.I. Kaikina, P.I. Naumkin, I.A. Shishmarev, Asymptotics for Dissipative Nonlinear Equations (2006)

Vol. 1885: A. Telcs, The Art of Random Walks (2006)

Vol. 1886: S. Takamura, Splitting Deformations of Degenerations of Complex Curves (2006)

Vol. 1887: K. Habermann, L. Habermann, Introduction to Symplectic Dirac Operators (2006)

Vol. 1888: J. van der Hoeven, Transseries and Real Differential Algebra (2006)

Vol. 1889: G. Osipenko, Dynamical Systems, Graphs, and Algorithms (2006)

Vol. 1890: M. Bunge, J. Funk, Singular Coverings of Toposes (2006)

Vol. 1891: J.B. Friedlander, D.R. Heath-Brown, H. Iwaniec, J. Kaczorowski, Analytic Number Theory, Cetraro, Italy, 2002. Editors: A. Perelli, C. Viola (2006)

Vol. 1892: A. Baddeley, I. Bárány, R. Schneider, W. Weil. Stochastic Geometry, Martina Franca, Italy, 2004. Editor: W. Weil (2007)

Vol. 1893: H. Hanßmann, Local and Semi-Local Bifurcations in Hamiltonian Dynamical Systems, Results and Examples (2007)

Vol. 1894: C.W. Groetsch, Stable Approximate Evaluation of Unbounded Operators (2007)

Vol. 1895: L. Molnár, Selected Preserver Problems on Algebraic Structures of Linear Operators and on Function Spaces (2007)

Vol. 1896: P. Massart, Concentration Inequalities and Model Selection, Ecole d'Été de Probabilités de Saint-Flour XXXIII-2003. Editor: J. Picard (2007)

Vol. 1897: R. Doney, Fluctuation Theory for Lévy Processes, Ecole d'Été de Probabilités de Saint-Flour XXXV-2005. Editor: J. Picard (2007)

Vol. 1898: H.R. Beyer, Beyond Partial Differential Equations, On linear and Quasi-Linear Abstract Hyperbolic Evolution Equations (2007)

Vol. 1899: Séminaire de Probabilités XL. Editors: C. Donati-Martin, M. Émery, A. Rouault, C. Stricker (2007)

Vol. 1900: E. Bolthausen, A. Bovier (Eds.), Spin Glasses (2007)

Vol. 1901: O. Wittenberg, Intersections de deux quadriques et pinceaux de courbes de genre 1, Intersections of Two Quadrics and Pencils of Curves of Genus 1 (2007)

Vol. 1902: A. Isaev, Lectures on the Automorphism Groups of Kobayashi-Hyperbolic Manifolds (2007)

Vol. 1903: G. Kresin, V. Maz'ya, Sharp Real-Part Theorems (2007)

Vol. 1904: P. Giesl, Construction of Global Lyapunov Functions Using Radial Basis Functions (2007)

Vol. 1905: C. Prévôt, M. Röckner, A Concise Course on Stochastic Partial Differential Equations (2007)

Vol. 1906: T. Schuster, The Method of Approximate Inverse: Theory and Applications (2007)

Vol. 1907: M. Rasmussen, Attractivity and Bifurcation for Nonautonomous Dynamical Systems (2007)

Vol. 1908: T.J. Lyons, M. Caruana, T. Lévy, Differential Equations Driven by Rough Paths, Ecole d'Été de Probabilités de Saint-Flour XXXIV-2004 (2007)

Vol. 1909: H. Akiyoshi, M. Sakuma, M. Wada, Y. Yamashita, Punctured Torus Groups and 2-Bridge Knot Groups (I) (2007)

Vol. 1910: V.D. Milman, G. Schechtman (Eds.), Geometric Aspects of Functional Analysis. Israel Seminar 2004-2005 (2007)

Vol. 1911: A. Bressan, D. Serre, M. Williams, K. Zumbrun, Hyperbolic Systems of Balance Laws. Lectures given at the C.I.M.E. Summer School held in Cetraro, Italy, July 14–21, 2003. Editor: P. Marcati (2007)

Vol. 1912: V. Berinde, Iterative Approximation of Fixed Points (2007)

Vol. 1913: J.E. Marsden, G. Misiołek, J.-P. Ortega, M. Perlmutter, T.S. Ratiu, Hamiltonian Reduction by Stages (2007)

Vol. 1914: G. Kutyniok, Affine Density in Wavelet Analysis (2007)

Vol. 1915: T. Bıyıkoğlu, J. Leydold, P.F. Stadler. Laplacian Eigenvectors of Graphs. Perron-Frobenius and Faber-Krahn Type Theorems (2007)

Vol. 1916: C. Villani, F. Rezakhanlou, Entropy Methods for the Boltzmann Equation. Editors: F. Golse, S. Olla (2008)

Vol. 1917: I. Veselić, Existence and Regularity Properties of the Integrated Density of States of Random Schrödinger (2008)

Vol. 1918: B. Roberts, R. Schmidt, Local Newforms for GSp(4) (2007)

Vol. 1919: R.A. Carmona, I. Ekeland, A. Kohatsu-Higa, J.-M. Lasry, P.-L. Lions, H. Pham, E. Taflin, Paris-Princeton Lectures on Mathematical Finance 2004. Editors: R.A. Carmona, E. Çinlar, I. Ekeland, E. Jouini, J.A. Scheinkman, N. Touzi (2007)

Vol. 1920: S.N. Evans, Probability and Real Trees. Ecole d'Été de Probabilités de Saint-Flour XXXV-2005 (2008)

Vol. 1921: J.P. Tian, Evolution Algebras and their Applications (2008)

Vol. 1922: A. Friedman (Ed.), Tutorials in Mathematical Biosciences IV. Evolution and Ecology (2008)

Vol. 1923: J.P.N. Bishwal, Parameter Estimation in Stochastic Differential Equations (2008)

Vol. 1924: M. Wilson, Weighted Littlewood-Paley Theory and Exponential-Square Integrability (2008)

Vol. 1925: M. du Sautoy, Zeta Functions of Groups and Rings (2008)

Vol. 1926: L. Barreira, C. Valls, Stability of Nonautonomous Differential Equations (2008)

Recent Reprints and New Editions

Vol. 1618: G. Pisier, Similarity Problems and Completely Bounded Maps. 1995 – 2nd exp. edition (2001)

Vol. 1629: J.D. Moore, Lectures on Seiberg-Witten Invariants. 1997 – 2nd edition (2001)

Vol. 1638: P. Vanhaecke, Integrable Systems in the realm of Algebraic Geometry. 1996 – 2nd edition (2001)

Vol. 1702: J. Ma, J. Yong, Forward-Backward Stochastic Differential Equations and their Applications. 1999 – Corr. 3rd printing (2007)

Vol. 830: J.A. Green, Polynomial Representations of GL_n, with an Appendix on Schensted Correspondence and Littelmann Paths by K. Erdmann, J.A. Green and M. Schocker 1980 – 2nd corr. and augmented edition (2007)

Printed in the United States
By Bookmasters